P9-DVJ-708

THE FOSSIL RECORD

Unearthing Nature's History of Life

THE FOSSIL RECORD

Unearthing Nature's History of Life

JOHN D. MORRIS

FRANK J. SHERWIN

Dallas, Texas
www.icr.org

The Fossil Record

Unearthing Nature's History of Life

by Dr. John D. Morris and Frank J. Sherwin

Dr. John Morris, perhaps best known for leading expeditions to Mt. Ararat in search of Noah's Ark, received his Doctorate in Geological Engineering at the University of Oklahoma in 1980. He served on the University of Oklahoma faculty before joining the Institute for Creation Research in 1984. Dr. Morris held the position of Professor of Geology and was appointed President in 1996. He continues to serve ICR as President Emeritus. He traveled widely around the world speaking at churches, conferences, schools, and scientific meetings. Dr. Morris has written numerous books and articles on the scientific evidence that supports the Bible. Dr. Morris is the author or co-author of such books as *The Young Earth, The Modern Creation Trilogy, The Global Flood: Unlocking Earth's Geologic History,* and *Noah's Ark: Adventures on Ararat.* He is also a contributor to *Guide to Creation Basics* and *Creation Basics & Beyond.*

Frank Sherwin received his bachelor's degree in biology from Western State College, Gunnison, Colorado, in 1978. He attended graduate school at the University of Northern Colorado, where he studied under the late Gerald D. Schmidt, one of the foremost parasitologists in America. In 1985, Mr. Sherwin obtained a master's degree in zoology. He published his research in the peer-reviewed *Journal of Parasitology*. He contributes his scientific expertise to a variety of ICR's publications on creation science and is one of ICR's most sought-after speakers. He is the author of *The Ocean Book* and *Guide to Animals*, co-author of *The Human Body: An Intelligent Design*, and a contributor to *Guide to Creation Basics* and *Creation Basics & Beyond.*

First printing: February 2010
Fifth printing: July 2017

Graphic Design: Susan Windsor

ISBN: 978-0-932766-98-4
Library of Congress Catalog Number: 2009940305

Please visit our website for other books and resources: ICR.org

Printed in the United States of America.

Contents

In memory of Henry M. Morris, Ph.D. (1918-2006), founder of the Institute for Creation Research

Acknowledgments

A book like this requires input from numerous individuals, and the authors would like to express their appreciation to several who played a significant part in bringing it to fruition. Dr. Duane Gish, who has long been the "fossil guy" at ICR, graciously reviewed the manuscript and wrote the Foreword. Fossil experts Dr. Kurt Wise and Dr. Gary Parker edited various stages of the book. Thanks too for the contributions of Joe Taylor, Bea Dunkel, John Doherty, and Brian Thomas, each providing valuable insights on the content. Many of the fossils that appear within these pages are from ICR's fossil collection, and several others were obtained from Heritage Auctions, Inc., often a source of beautiful fossils. As noted in the Photo and Illustration Credits, numerous colleagues have kindly allowed the use of their photographs and images.

Special gratitude must be expressed to the professional publishing staff at ICR, including Managing Editor Beth Mull, Assistant Editor Christine Dao, Associate Book Editor Dr. Brad Forlow, and special thanks to graphic designer Susan Windsor for preparing the beautiful layout of the book and many of its fine illustrations.

Foreword

Dr. John Morris received his Ph.D. in geological engineering from the University of Oklahoma and is President of the Institute for Creation Research in Dallas, Texas. Frank Sherwin received his graduate training in biology from the University of Northern Colorado. Both have extensive resumes in creation research and ministry. During their many travels throughout the United States and abroad, they have lectured in schools, universities, churches, and other venues, presenting the scientific and biblical evidence for creation. The combination of Morris' training and experience in geology and Sherwin's expertise in zoology has resulted in this beautiful and persuasive book on *The Fossil Record.*

There are some evolutionists who, while defending evolution, do admit that the fossil record provides adequate evidence to determine which has the strongest evidence, creation or evolution. Thus, evolutionists Glenister and Witzke state that "the fossil record affords an opportunity to choose between evolutionary and creationist models for the origin of the earth and its life forms."[1]

Futuyma expressed a similar belief when he said:

> Creation and evolution, between them, exhaust the possible explanations for the origin of living things. Organisms either appeared on the earth fully developed or they did not. If they did not, they must have developed from pre-existing species by some process of modification. If they did appear in a fully developed state, they must have been created by some omnipotent intelligence.[2]

The authors begin by documenting that the worldviews of those involved in this contest are of considerable importance. Thus, Richard Lewontin, evolutionist and Harvard professor, states:

> Yet, whatever our understanding of the social struggle that gives rise to creationism, whatever the desire to reconcile science and religion may be, there is no escape from the fundamental contradiction between evolution and creationism. They are irreconcilable world views.[3]

The authors make it abundantly clear that theirs is the biblical worldview, which holds that God created the earth and all living organisms, as related in the first ten chapters of the book of Genesis. They then describe what should be found in the fossil record if creation is true and contrast that to what should be found if evolution is true. This discussion includes information on important aspects of geology and dating methods.

As you read this book, you will find that Morris and Sherwin indeed present powerful scientific evidence from the fossil record that living organisms appeared abruptly on the earth in a fully formed state and remained in stasis until the present day. They exhibit amazing complexity from the start, not simple-to-complex evolution. Their fossils point to rapid burial in catastrophic, watery conditions, and are typically found today in mass fossil graveyards with organisms from mixed habitats, having suffered agonizing deaths. These features are just what should be present if they resulted from creation and a cataclysmic flood such as that recorded in the Bible.

The authors then dig into the most interesting part of this book as they lay out the scientific evidence from the fossil record that in many cases is devastating to evolution. One example was so utterly contradictory to evolution that the evolutionist source proclaimed that "this is one count in the creationist's charge that can only evoke in unison from the [evolutionary] paleontologists a plea of *nolo contendere*"![4] Thus, they leave the evolutionists in a position of no defense.

Morris and Sherwin have thoroughly searched the scientific literature that augments their personal field study of the fossil evidence related to many features from fossils of the smallest organisms to the origin of man. They have assembled a wealth of material that is sufficient to enable all who study the fossil record with an open mind to realize that the record declares that "in the beginning God created the heaven and the earth." I urge everyone who has an interest in the fossil record to obtain a copy of this excellent book.

Duane Gish, Ph.D.
Senior Vice President Emeritus
Institute for Creation Research

The name *Sinosauropteryx prima* means “first Chinese lizard-wing.” This specimen was found in the famous Jiufotang Formation in central Asia, supposedly deposited about 100 million years ago, from which numerous outstanding fossils have been recovered. Tiny grains dominate the deposit, allowing even ephemeral features to be preserved. The genus to which the dinosaur belonged has also yielded similar fossils that seem to sport feathers, which some scientists claim supports the evolutionary idea that dinosaurs evolved into birds. This position, however, is not held by all scholars. In this fossil, no feather impressions are visible. Note the fish fossil nearby. This type of dinosaur did not live in the water, but both were encased in watery sediments. The fish was caught writhing in suffocation, and the dinosaur was gasping for breath.

Introduction

The creation/evolution controversy shows no sign of letting up. Although the majority of individuals on both sides nominally believe what they were taught in school or at home, many have become aware that this is a seminal issue, perhaps the most important of our day. They see it as a worldview battleground, one that cannot be ignored.

Evolution, which purports to account for all things through purely natural processes, gives seeming credibility to the naturalistic worldview. Unless one is omniscient, however, one could not possibly know there is nothing supernatural. Wouldn't that itself be a claim for godhood? Evolutionists make the assumption of naturalism, and choose to approach the scientific enterprise—and often, their life decisions and destinies—as if there were no God. Or, if He does exist, as though He has never acted in time and space.

Creation presents a supernatural explanation for the origin of all things, and it gives the theistic worldview its foundation. Christianity claims to be rooted in fact, and holds that the record of Scripture communicates scientific and historical truth. It may not give all the details, but it does claim that the events recorded there actually happened and that these events make provision for an eternal relationship between created individuals and their Creator. The existence of God is as much a faith assumption as naturalism, but creationists insist that it is a more reasonable faith, buttressed by much observation of scientific fact.

Historians have long recognized that Charles Darwin began his career as a creationist, studying for the ministry. Although he loosely accepted the creation account of Genesis 1 and 2, he came to doubt the ruination of creation due to man's rejection of his Creator, as recorded in Genesis 3. Thus, Darwin had no way to account for the presence of pain, suffering, and death (the penalty for mankind's rebellion) in a supposedly "very good" creation. Personal suffering led him to choose to believe that there was no supernatural God and that only natural processes were at work in the world. He therefore set about to find a naturalistic cause for the living creatures he observed.

Natural selection was his conclusion. Darwin became an aggressive advocate of religious naturalism, and today many are following in his footsteps of faith. Scripture and its creation doctrine have always been the same, but most people in the Western world have never had the chance to learn creation thinking and know only evolution. Naturalism enjoys a virtual monopoly in today's classrooms, while instructors who have been schooled only in a naturalistic worldview play the part of evolutionary evangelists.

Less well-known is the fact that many leading anti-creationists of today come from Christian backgrounds. Many even testify to having gone to "Christian" churches, schools, and universities, where they were taught evolutionary naturalism in the name of science. They abandoned their Christian faith, adopted another belief system, and now ardently "preach" this different faith.

A leading evolutionist, now teaching at Harvard University, gave his "testimony" in a naturalistic journal:

> As were many persons from Alabama, I was a born-again Christian. When I was fifteen, I entered the Southern Baptist Church with great fervor and interest in the fundamentalist religion; I left at seventeen when I got to the University of Alabama and heard about evolutionary theory.[1]

Students in college may leave the Christian faith for many reasons, but those who try to give a systematic explanation routinely cite evolution as the cause. The most successful "evangelistic" tool for naturalism throughout the years has no doubt been an evolutionary presentation of the fossil record. For instance, in the Petrified Forest in Yellowstone National Park, many petrified trees stand upright in numerous layers of rock, which are claimed to hold successive forests. The total growth time required for all the forests in succession

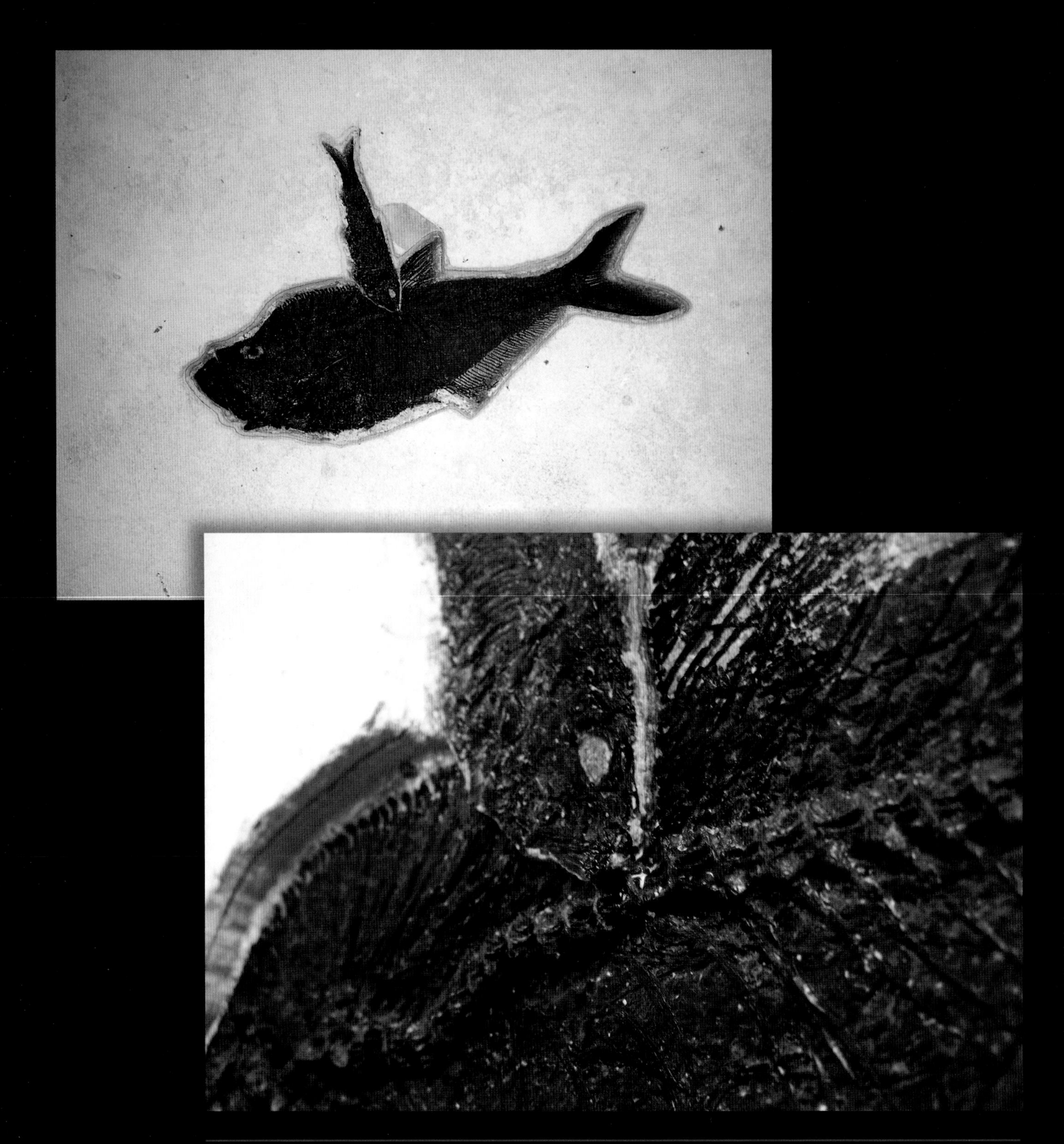

The formation in which these overlapping fish were entombed contains myriads of laminae. Uniformitarians believe that each tiny pair of layers represents one year's worth of sedimentation in a calm environment. The main fish was smashed flat, yet no distortion or decay can be seen. This seemingly denies the "one layer per year" idea. At the very least, it refutes the principle of uniformity. (Location: Green River Formation, Wyoming. Dated as Eocene.)

exceeds many thousands of years. Old-earth advocates often use this as the best proof that Scripture contains error, since they claim the earth could not be as young as the Bible says. However, although the trees are in growth position, these advocates fail to ask the crucial question—are they in growth location?

One leading anti-creationist, formerly a Christian/creationist, cites this evidence as his primary reason for abandoning Christianity and adopting evolution:

> I vividly remember the evening I attended an illustrated lecture on the famous sequence of fossil forests in Yellowstone National Park... first agonizing over, then finally accepting, the disturbing likelihood that the earth was at least thirty thousand years old. Having thus decided to follow science rather than the Scripture on the subject of origins, I quickly, though not painlessly, slid down the proverbial slope toward unbelief.[2]

We now know the information he received was quite incorrect. The trees are not standing forests at all, but were trapped in successive mudflows, with a few of them maintaining an erect posture. Matching tree-ring patterns from several layers prove they all grew at the same time.[3] Thus, there was no reason to abandon Scripture.

This book desires to counter such evolutionary misinformation, provide a more accurate look at the fossils, and stop the hemorrhaging of the Christian faith. Fossils have for too long been evolutionists' favorite weapon in the creation/evolution battle, with far too many casualties. The church has lost enough of its young people to wrong thinking, leading to our desire to remove fossils from the evolutionary arsenal.

Typically, students receive only a one-sided presentation of the fossil record. But the claim that fossils document evolution is simply not true. The fossil record records a very different message, one supportive of the creation worldview. It speaks of exquisite design in every once-living thing, not random development solely through natural processes. Each created kind appeared abruptly, rather than being altered from a previous kind. The fossils show variation and adaptation, but not evolution. There is no hint of an evolutionary "simple to complex" history, for life was complex from the very start. Instead, the overwhelming message of the fossil record is one of stasis, not evolutionary change. The fossils often give testimony to violence at the time of their death and burial. All these testify to the biblical history of recent creation, the Curse due to Adam's sin, and the great Flood of Noah's day.

An evolutionary overprint laid over the fossils holds power only if the alternative is censored. But censorship is not truth; there is a better way to think. Adopting evolutionary naturalism as one's faith and guideline for life makes no sense if there is a God who has spoken. This book can help you choose your faith and discover the Creator of all things. And for those who wish to share the real nature of the fossil record, selected images from the book are available online at icr.org/fossil-record for use in PowerPoint presentations.

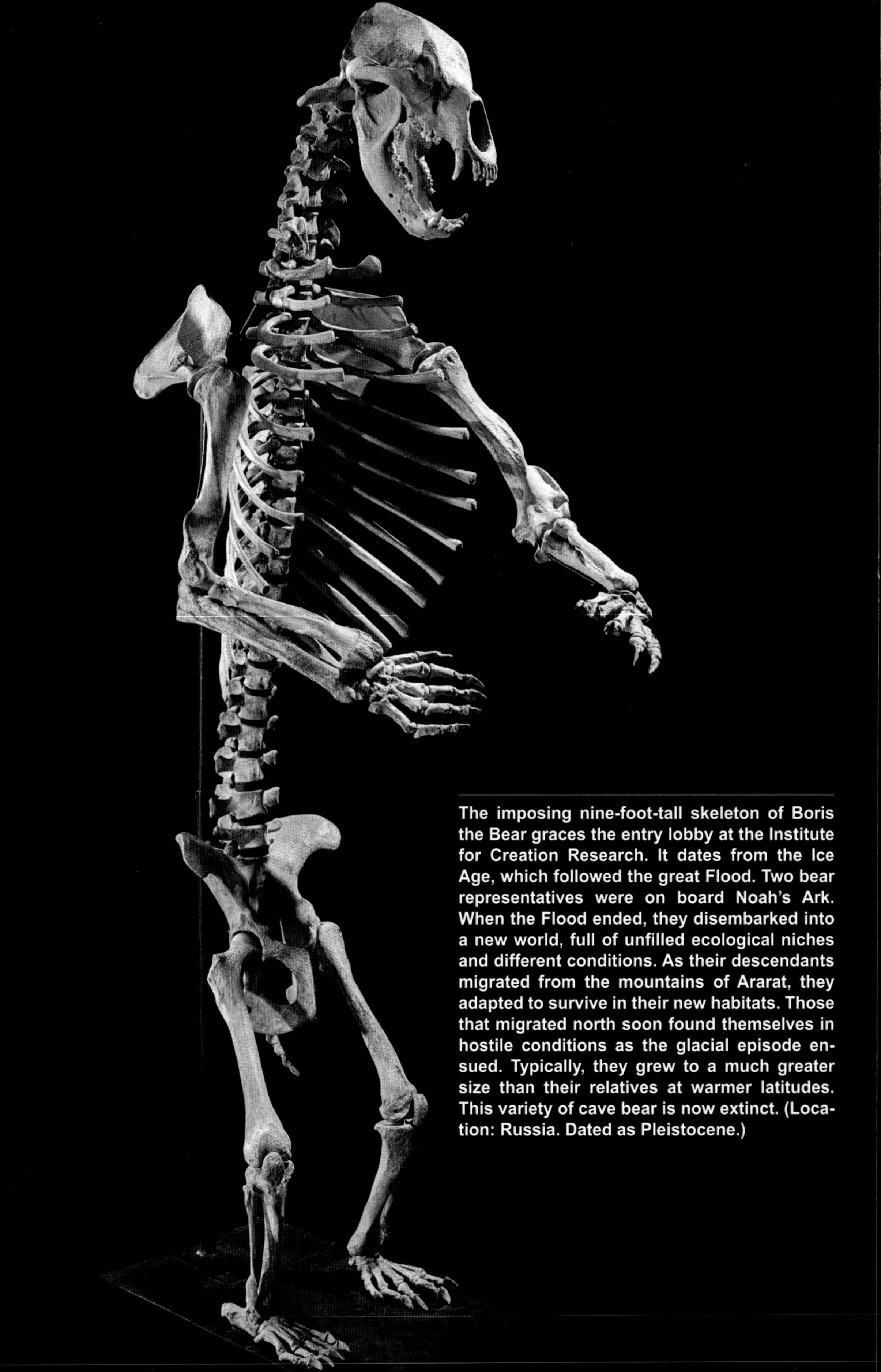

The imposing nine-foot-tall skeleton of Boris the Bear graces the entry lobby at the Institute for Creation Research. It dates from the Ice Age, which followed the great Flood. Two bear representatives were on board Noah's Ark. When the Flood ended, they disembarked into a new world, full of unfilled ecological niches and different conditions. As their descendants migrated from the mountains of Ararat, they adapted to survive in their new habitats. Those that migrated north soon found themselves in hostile conditions as the glacial episode ensued. Typically, they grew to a much greater size than their relatives at warmer latitudes. This variety of cave bear is now extinct. (Location: Russia. Dated as Pleistocene.)

Chapter 1 Evolution and Creation as Worldviews

Have you ever been asked, "Do you believe in evolution?" If you're talking to your teacher or professor, the expected, perhaps demanded, answer is "yes." If you even waver, it could jeopardize your final grade. After all, everybody believes in evolution, don't they? Isn't that what everyone has been taught? Don't only ignorant yokels have any doubts about its truth?

But obviously, not everybody "believes" in evolution. Polls consistently reveal that a great majority of Americans do not believe that all living things evolved from a common ancestor. Why isn't evolution something that can be known for sure, not just something that must be believed?

It helps to define terms before that question is asked. These terms are often loosely applied, but the following definitions are the ones that will be used in this book. They mirror common usage, but not always technical usage.

First *evolution*, as it is used here, is the "descent from a common ancestor" model, which holds that all of life developed from more primitive forms. Starting with single-cell organisms, life supposedly followed a chain of development from marine invertebrates, to chordates, to fish, to early reptiles and amphibians, to various stages of mammals, and finally to the pinnacle of evolutionary development—humans. This colossal chain of change was all brought about by time, chance, struggle, and death, without help from any form of a supernatural "god" or an "intelligent designer." Mutation and natural selection did it all.

It has not really dawned on most people that human evolution entails such big changes. They might have made a nominal peace with the concept that man came from the apes, but have they considered that man would have also had to come from insectivores and invertebrates? A revealing question to ask them might be, "Do you believe your ancestors were fish, and before that a marine invertebrate, as evolution teaches?" Not many people would insist on answering "yes"! Despite several generations of aggressive evolution-only teaching in the public school classroom, most people know instinctively that they did not come from an insectivore or a fish or a starfish. People can choose to believe they have such an animal ancestry, but few really do accept it in their hearts. It just is not credible.

Thankfully, evolution is not the only alternative for origins. The other option is not only more believable, it is more appealing. And, as we shall see, it is also more scientifically credible.

Another issue arises here. For many people, evolution carries with it an implied belief in atheism. If all things came about through natural processes alone, there is no need for a supernatural Creator and no God to whom man is ultimately accountable. Evolution can better be understood as the pseudo-scientific justification for a life lived without accountability to one's Maker. Evolution has become state-sponsored religious dogma, dictating belief not only on the subject of origins, but decreeing a complete, exclusive worldview.

Dr. Michael Ruse, thought by some to be the chief spokesman for evolution today, has admitted:

> Evolution is promoted by its practitioners as more than mere science. Evolution is promulgated as an ideology, a secular religion—a full-fledged alternative to Christianity, with meaning and morality. I am an ardent evolutionist and an ex-Christian, but I must admit that in this one complaint...the literalists are absolutely right. Evolution is a religion. This was true of evolution in the beginning, and it is true of evolution still today.[1]

Christianity recognizes the existence and supremacy of God. As a worldview, it answers the questions of who we are, where we came from, the meaning of life, and where we're going after we die. It tells us how to live and how to make life decisions along the way. The evolutionary worldview also addresses these subjects, but with different answers to the same questions.

Christianity claims that humans are created in the image of God. We have immense worth in His eyes, and a high standing before Him when we appropriate His gift of redemption. We have been granted a great destiny to perform while on earth, and life with our loving and righteous Creator/Savior after death.

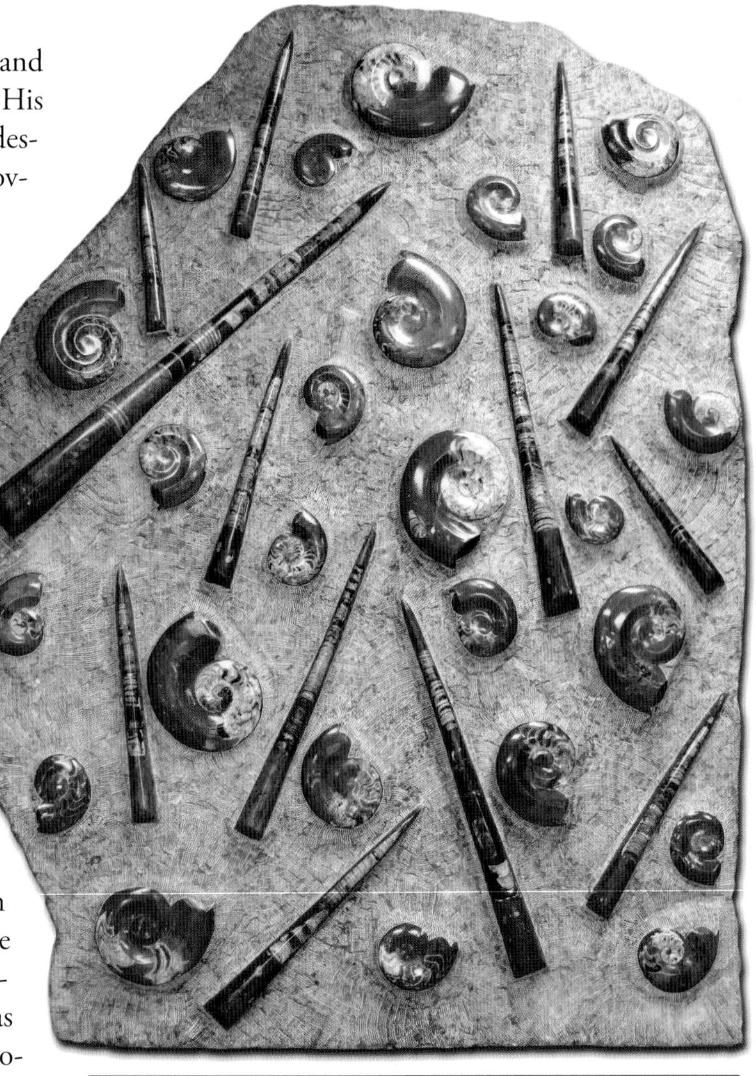

This ten-foot by six-foot fossil slab of limestone comes from a rich fossil site in Morocco, the source of many specimens purchased in rock shops around America. The fossils are calcified and can be polished. This slab contains at least two varieties of cephalopods. These squid-like, ocean-dwelling creatures typically had a shell. Some were coiled, some were straight, but some had no shell at all. They were strong swimmers and lived in different environments, so we wonder how they were buried together. It must have required unusual, dynamic conditions, such as a rapid underwater gravity flow of sediments. Note the preferred direction of the long axis of the animals, with the pointed ends facing the current. This required moving water laden with sediments.

According to evolution, however, humans come from the universe's chemicals, which were self-organized through purely natural processes into unlikely forms over eons of time. Single-cell life transformed itself into higher forms, until finally the human animal emerged. As higher animals, we have incorporated animal behavior into societal norms. The only true meaning to life is survival and reproduction, and life's highest goal is to pass on one's genes more efficiently than others do. At death, we simply cease to exist.

Yes, as Dr. Ruse explains, "evolution is a religion" and not science. It might be best understood as a way of thinking about and interpreting the world around us. Some have called this worldview "philosophical materialism"—a religious claim of naturalism, which asserts that nature is all there is. There is no supernatural Being who has interfered with the natural order of things. Perhaps, as creation biologist Dr. Gary Parker often claims, evolution is just "humanism dressed up in a lab coat," a statement of belief regarding all of reality.

Evolution is certainly not a science in the traditional sense of empirical science, which employs the scientific method of observation and testing. The word "science" comes from the Latin *scienta* (knowledge). Does evolution fit in this category? If life ever sprang from non-life, or if one kind of plant or animal ever spontaneously developed into a different kind, it did so in the unobserved past. These processes are simply not observed happening today. How can evolution then be considered knowledge? How can it be known to have happened in the past, when no one was present to observe it? If speculation regarding possible unobserved events must be called "science," let it be called "historical science," or "forensic science," or "origins science" to differentiate it from observational science.

In a conceptual sense, creation thinking about history exactly parallels evolutionary thinking. Neither creation nor evolution can be directly observed. Both reconstructions of history strictly adhere to natural law operating in the present. Both viewpoints see the universe and all within it operating through natural processes, and preclude supernatural intervention on an ongoing basis.

When it comes to origins and earth history, however, there is a very important difference between creation and evolution. Evolutionists insist there were no observers to pre-human history and that no reliable records exist to consult. Christian creationists, on the other hand, claim that the Bible is an absolutely accurate record of the acts of the Creator in real history. If evolutionists are free to build their beliefs on the words of Darwin and others who certainly were not present to observe the events of early earth history, then surely creationists are free to consult and build upon the written record of the Ultimate Observer who made that history happen.

But real science is not of necessity purely naturalistic, either; science is a search for truth about nature. It is not a search for "naturalistic explanations" no matter what the evidence suggests. "Nature" produces patterns that reflect time, chance, and the properties of matter. Only an overriding mind can produce patterns that involve plan, purpose, and organization, all of which reflect the designer's goals.

Creation requires supernatural input in past origins events that were unobserved by man, and thus is not strictly naturalistic. Conversely, evolution denies the need for the supernatural, which is an equally "religious" concept. Some evolutionists do claim a belief in God, but evolution itself needs no God, and nearly all founders and leaders of evolutionary thinking hold no such belief. They know there is no room in evolution for the supernatural. Even "theistic evolution" (the idea that God used evolution as His method of creation) is indistinguishable from non-theistic evolution as it pertains to the interpretation of the historical data.

Evolutionary adherents demand that the science classroom be dominated by this perspective. Consider this quote from Richard Lewontin, a leading evolutionist:

> We take the side of science in spite of the patent absurdity of some of its constructs, in spite of its failure to fulfill many of its extravagant promises of health and life, in spite of the tolerance of the scientific community for unsubstantiated just-so stories, because we have a prior commitment to materialism....We are forced by our *a priori* adherence to material causes to create an apparatus of investigation and a set of concepts that produce material explanations, no matter how counterintuitive, no matter how mystifying to the uninitiated. Moreover, that materialism is absolute, for we cannot allow a Divine Foot in the door.[2]

Dr. Lewontin's statement demonstrates the absolute stranglehold materialistic atheism has on every thought or theory that is allowed to be considered in the scientific and educational realms. This makes the American classroom one of the most censored, thought-controlled locations on the planet.

While we cannot go back in time to the very beginning and make scientific observations about origins events, we can and must do better science than what is demanded by naturalists. We can study the presently existing evidence that resulted from past events, and compare how well each worldview explains the data. The view that more correctly and consistently coordinates the data is the one that is more likely to be accurate.

While this book is titled *The Fossil Record*, the following pages will not examine all fossils, as would a technical treatment of the subject. It will instead limit itself to the creation/evolution question and how each perspective views the fossil data. We will attempt to determine which worldview best handles the evidence, and therefore which one is most reasonable and most likely correct.

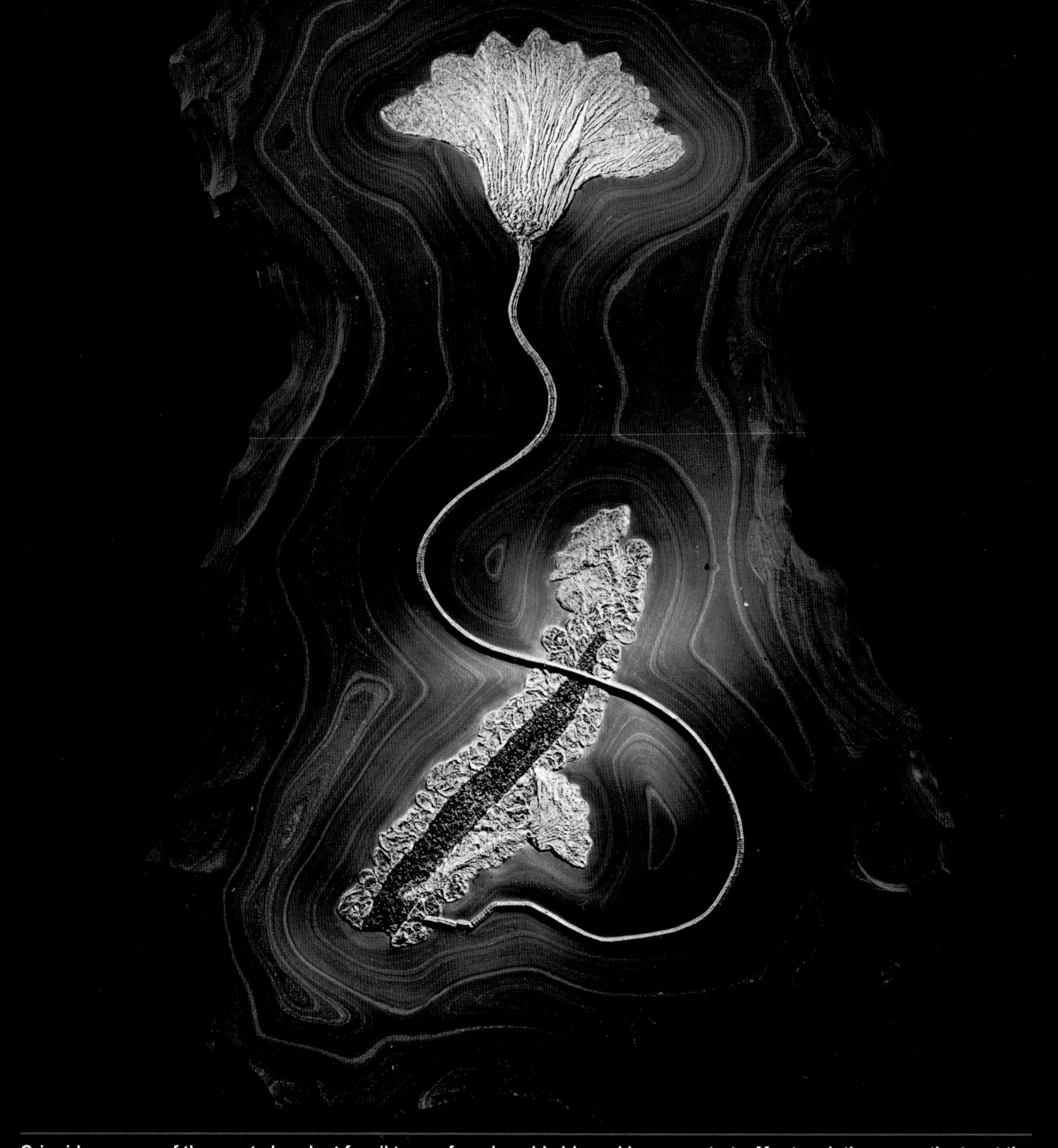

Crinoids are one of the most abundant fossil types, found worldwide and in many strata. Most varieties are extinct, yet the modern sea lily can be counted as a "living fossil." Despite its plant-like appearance, it is classed as an animal, feeding on nutrients in the surrounding seawater. Attached to the ocean bottom during life, upon death its stalk almost immediately disintegrates into thin disks. This specimen remained intact as it was buried, likely in rather calm water. Although the great Flood of Noah's day can rightly be termed a catastrophe, certain zones were quiet as fossils were entombed in settling sediments. This animal was attached to floating wood. (Location: Germany. Dated as Lower Jurassic.)

Chapter 2 Creation/Evolution Battle Starts Afresh

Most observers maintain that the modern revival of creation thinking began in 1961 with the publication of *The Genesis Flood* by John Whitcomb and Henry Morris. Few scientifically minded individuals held to the biblical view of earth history at that time. Some preachers did, but they essentially held this view without supporting scientific evidence. Naturalistic evolution was generally taught as fact, with no dissenting opinion.

Those who doubted Darwin's theory typically held that the Bible taught "fixity of species," the idea that species did not change and that modern species were the very ones God had created in the beginning, except for those that had since gone extinct. But this was an untenable position. The concept of a "species" may have been incompletely defined, but every biologist knows of species that vary and populations that change over time, sometimes leading to new species. Thus, to a trained biologist, fixity of species was an inadequate scientific explanation for what can be observed.

Eventually, several brave Christians familiar with biology began to ponder the definition of species and the limits of biological change. Change happens, but how much change can occur? Do the limited changes that are readily apparent add up to the extensive changes required by evolution, in which fundamentally different body styles arise from others?

Creationist speculation regarding the extent of variation centered on the meaning of the Hebrew phrase "after its kind," which is used 10 times in Genesis 1 to describe the creation of the various groups of living organisms. What is a "kind"? Was this the basic creation unit? Most professionals eventually concluded that limited lateral variation was in keeping with creationist tenets. Once it was acknowledged that some variety was acceptable within the limits of the "kind," biologists were free to recognize change without necessarily having to embrace the concept of evolutionary change. It became possible to be an intellectually honest biologist without accepting the theory of evolutionary descent from a common ancestor.

Interestingly, limited change is all that scientists of any viewpoint have ever observed. This is fully compatible with biblical teaching. Variety is all around us, exhibiting abundant "horizontal" change within distinct groups. But variety is not evolution, which necessitates extensive "vertical" change. Mutations frequently occur, but random mutations do not produce the constructive changes needed by evolution. The same could be said for natural selection, novel genetic recombinations, etc. Whenever an evolutionist cites evidence for "evolutionary" change, he cites only evidence for variety within a created kind (or *micro*evolution), not the origin of new kinds (or *macro*evolution). Fossils documenting minor transitions within a "kind" are abundant, while fossil links between kinds are systematically missing. There seems to be no fossil record of evolution on the large scale needed.

One of the most effective creation books in those early days was the little book by Dr. Duane Gish entitled *Evolution: The Fossils Say No!* Written in non-technical language, yet presenting meaty considerations of fossils typically used in support of evolution, it had an enormous impact. Revised several times, it eventually became *Evolution: The Challenge of the Fossil Record* and *Evolution: The Fossils Still Say No!* These more technical and documented treatments became public enemy number one to the evolutionist.

You see, if you asked an evolutionist in the 1960s or '70s, "Where is the evidence for evolution?," the answer would likely have been, "In the fossil record, the record of life in the past." But when asked to produce evidence of past body-style transitions between ancestor and descendant, they could demonstrate variety within limits, but no transitional forms between basic kinds. This became a serious embarrassment to evolutionists.

In the 1970s, the new evolutionary concept of "punctuated equilibrium" acknowledged the obvious fossil gaps. It proposed a more rapid pace for evolution, one so rapid and sporadic that it would leave no "in-between" fossils. This idea was enthusiastically received by students and professors alike, since it seemed to explain the lack of fossil transitions. However, it left careful thinkers unsatisfied, for it only dealt with minor alterations between closely related species, and not the larger gaps between fundamentally different kinds. Primarily an argument from silence, it explained only the lack of important fossil evidence. It also offered no adequate biological mechanism to accomplish the rapid major changes needed for the theory to work.

The late Harvard paleontologist and devout evolutionist Dr. Stephen J. Gould was the leading proponent of punctuated equilibrium. A prolific author, he penned many impassioned arguments for rapid "punctuated" evolution, and admitted the paucity of evidence for major change:

> The history of most fossil species includes two features particularly inconsistent with gradualism: 1) Stasis—most species exhibit no directional change during their tenure on earth. They appear in the fossil record looking much the same as when they disappear; morphological change is usually limited and directionless; 2) Sudden appearance—in any local area, a species does not arise gradually by the steady transformation of its ancestors; it appears all at once and "fully formed."[1]

Later, he admitted he could not even imagine what the intermediate forms would look like:

> All paleontologists know that the fossil record contains precious little in the way of intermediate forms; transitions between major groups are characteristically abrupt.
>
> Even though we have no direct evidence for smooth transitions, can we invent a reasonable sequence of intermediate forms, that is, viable functioning organisms, between ancestors and descendants? Of what possible use are the imperfect incipient stages of useful structures? What good is half a jaw or half a wing?[2]

One of Gould's primary collaborators in the new theory was Dr. Steven Stanley, who noted:

> The known fossil record fails to document a single example of phyletic evolution [i.e., involving a basic body style change] accomplishing a major morphological transition and hence offers no evidence that the gradualistic model can be valid.[3]

A more recent colleague insists the stage has not been reset, and the call for rapid, undocumented spurts of evolution has not abated.

> Evolution seems to have happened in fits and starts—at least that's what the fossil record shows. From trilobites to pterodactyls, ammonites to Archaeopteryx, scientists find the same pattern: brief bursts of innovation in which a single species or branch on the tree of life turns into a cluster of new twigs, then lapses into long stretches ruled by the status quo.[4]

While paleontologists who dealt with the lack of transitions quickly accepted punctuated equilibrium, geneticists did not. They could not conceive of any genetic way to accomplish rapid transitions of coordinated soft and hard parts. The argument raged for several years, but seemed to vanish in a truce. Evolutionists in general agreed not to use the fossil record as evidence for evolution, for it did not seem to point that way. Instead, evidence from genomics and other new fields has taken top billing.

> No real evolutionist uses the fossil record as evidence in favor of evolution over creation.[5]

One thing is certain. If abundant transitional fossils existed, they would be trumpeted far and wide as proof of evolution. No recourse to hypothetical "missing links" would be necessary, and no recourse to rapid evolutionary leaps would either.

> Major transitions in biological evolution show the same pattern of sudden emergence of diverse forms at a new level of complexity. The relationships between major groups within an emergent new class of biological entities are hard to decipher and do not seem to fit the tree pattern that, following Darwin's original proposal, remains the dominant description of biological evolution. The cases in point include the origin of complex RNA molecules and protein folds; major groups of viruses; archaea and bacteria, and the principal lineages within each of these prokaryotic domains; eukaryotic supergroups; and animal phyla. In each of these pivotal nexuses in life's history, the principal "types" seem to appear rapidly and fully equipped with the signature features of the respective new level of biological organization.[6]

By the way, do you know what a missing link is? If not,

don't feel bad. Nobody does—because they're missing! They do not exist or have not been found! How could you possibly know what a missing link is if it is missing? Evolution requires untold billions of transitional forms, yet few fossil missing links are claimed to have been discovered, and they too are controversial, even among evolutionists. Why is this? Maybe it is because these missing transitional fossils never existed, as attested by this recent proponent: "The origin of animals is almost as much a mystery as the origin of life itself."[7]

Darwinian scholar Richard Milner is quoted as saying that Stephen J. Gould "took issue with those who used natural selection carelessly as a mantra, as in the evidence-free 'just-so stories' concocted out of thin air by mentally lazy adaptationists."[8] Gould himself wrote, "Our technical literature contains many facile verbal arguments—little more than plausible 'just-so' stories."[9]

This book presents evidence that actually exists. Our contention is that the fossil record's "big picture" does not support an evolutionary view of the past at all. Rather, it is fully compatible with the biblical teaching that in the beginning God created the various kinds of plants and animals, and allowed them to vary within limits over time. The fossil record speaks eloquently of creation, not evolution.

The "tree of life" was the only illustration Darwin used in his famous book. More than any other, it embodies the evolutionary mindset that all of life came from a common ancestor. The origin of life from non-life, or "spontaneous generation," is so unlikely it could hardly happen once, let alone multiple times. Darwin believed it happened only once and the life form produced descended into all others. Evolutionists, however, no longer feel it represents reality. Current thinking postulates that life may have arisen multiple times, or that things branched off into an "orchard" or a complex "web of life" rather than a "tree." An orchard analogy for fossils better speaks of creation than evolution. (See page 44.)

Chapter 3 Evolution and Creation Contrasted

It should be obvious that only two broad models of origins are possible. Either the universe and all life within it are the result of recognized forces—the same natural ones observed today—or its present makeup requires other processes, forces that do not operate today. Certainly, shades of opinion reside among proponents within each basic model, but couched this way, there is no other possibility. Either the universe is self-made and self-transforming, or it was produced by transcendent influences—i.e., forces and/or agencies outside the physical universe.

Neither creation nor evolution is in the realm of observation. Since they offer differing ideas about events in the unobserved past, it is more proper to call them scientific "models" of history and formally compare how well each meshes with the facts. Given that no view of history can be known through scientific observation to be absolutely true, what can we do? How can we conduct a meaningful study of the historical data, the fossils? One way is to compare the "predictions" of both models as they relate to the fossils, or the expectations of the fossil data that emanate from each. In doing so, we will attempt to discern which model handles the data better and is therefore more likely to be correct.

Several lines of reasoning bear on the overall origins question, but in this book we will confine our investigation to the fossil data. We will test ideas about the past against evidence from natural processes operating today.

Keep in mind that every scientist, whether creationist or evolutionist, agrees to limit their scientific exercise to only the natural processes operating today. Supernatural processes seldom, if ever, operate in the present, but were they needed in the past to produce what we see today? Patterns of plan and purpose versus patterns of time and chance can be recognized and investigated scientifically. How do we best account for what is observed? Creation scientists do not allow themselves to invoke miracles to explain present processes. It is crucial for all scientists to differentiate between "origin" events and "operational" events.

Evolutionists insist that natural processes have been "uniform" throughout history, having been governed by the same laws operating at the rates, scales, and intensities possible today. Obviously, this view of the past cannot be known with certainty, but it does employ elegant logic. "Uniformitarianism" involves a strict adherence to "uniform" processes that have acted consistently throughout the remote past. This principle disallows not only miraculous processes, but also large-scale, catastrophic processes. Events of the unseen, unknowable past are interpreted within the constraints of today's norms.

Creationists agree in general regarding present natural processes, but deny evolutionary naturalism and uniformitarianism as mandates for understanding how the current structure of the earth and its inhabitants came to be. To a creationist, the way processes operated in the past and the way they operate today must be very different. Something more than today's slow, gradual, local geologic processes were required to blanket the continents with multiple layers of fossil-rich deposits. Creationists claim natural law and the natural processes we study today are not capable of producing the current high state of order and level of energy. Something else, something beyond natural processes, was needed to originate the dazzling diversity of organisms that are well-designed to thrive in varied habitats, and the wide

Ammonites are common in the fossil record. The rare, loosely coiled ones are the most interesting; they are sometimes even quite bizarre. As cephalopods, they are related to the modern squid and Chambered Nautilus, and gain a section of length during each growth period. They swam and floated in water to find their food. They were also prey for larger animals, such as whales and sharks. Changes among ammonites have been used by evolutionists as evidence of evolutionary change. Creationists freely acknowledge variety within an animal group, but recognize that variety is not evolution. God designed each "kind" to vary and survive in whatever conditions they found themselves. Such "microevolution" should not be mistaken for evolution. (Location: Pierre Shale, South Dakota. Dated at 70 million years.)

array of geologic deposits that required processes operating at rates, scales, and intensities only hinted at today.

The Evolution Scenario

As commonly taught, the evolutionary worldview assumes naturalism, which holds that all things were built through strictly natural processes. In general, evolutionary naturalism starts with the simple and progresses toward the more complex, from an ethereal quantum nothingness to a cosmic egg, and then the Big Bang. Subatomic particles appeared and then proceeded to combine into hydrogen and helium, which over time developed into stars and galaxies. Some stars underwent supernovae, throwing out larger and more complex elements. Leftover stardust coalesced into orbiting planets and moons. At least one solar system witnessed the spontaneous generation of life on one of its planets, resulting in life forms. Prokaryotic life came first, cells that had no nuclear membranes around their DNA and yet were still immeasurably more complex than non-life. These cells evolved into eukaryotes, which had nuclear and other internal membranes, and these over time developed into multi-cellular life that eventually split into plants and animals.

Plants developed spores for reproduction, and then seeds. Flowering plants later adorned the planet's surface. Animal life took a different, but parallel, path. Single-cell life became the multitudes of marine invertebrate species, at least one of which in time developed into vertebrates (animals with spines). Jawless fish developed bone and filled the seas. Eventually, some emerged on to land as amphibians and then became reptiles. From them came the birds and mammals. Only at the very end of a long, long trail did some mammals become self-aware. As thinking beings, they developed language, civilization, and technology.

Conversely, the creation model assumes an adequate, supernatural Force capable of creating all things. This is faith, but not a blind faith, since the creation bears His signature. In the beginning, this almighty God created all of reality, time, space, matter/energy, and light, including the planet earth. He then shaped and fitted it for the life He would create next. He made plant life for the newly formed continents. Stellar bodies followed, providing an ongoing energy source. Next, animal life appeared in great abundance and variety. Finally, mankind—male and female—was created.

Either our origin was accomplished by natural processes without assistance from an intelligent designer, or it required a supernatural source. A leading evolutionary biologist acknowledged as much in a dogmatic anti-creationist book, in which he claimed:

> Creation and evolution, between them, exhaust the possible explanations for the origin of living things. Organisms either appeared on the earth fully developed or they did not. If they did not, they must have developed from pre-existing species by some process of modification. If they did appear in a fully developed state, they must indeed have been created by some omnipotent intelligence.[1]

The evolutionary concept of "developing from pre-existing species" needs some commentary. At each stage along the way, new traits and body parts would have been needed to continue the progression. The organism itself would not consciously know what was needed, nor could it plan the necessary mutations to bring about the changes. In hindsight, however, evolutionists recognize the need and see branching nodes on the so-called tree of life where they believe species diverged. Of course, since all of life's development and operation, as well as the reproduction of that life, require appropriate genetic instructions, each alteration would have necessitated new genes. For instance, a fish normally has genes for neither legs nor lungs, and must acquire them before it can make the transition from sea to land.

Genetic information is ingeniously simple in concept, and yet unimaginably complex in operation. If it could be randomly acquired, each acquisition would require a lengthy trial and error process for natural selection to weed out all harmful traits. In evolution, no intelligence is present to purposefully write the necessary genetic "software." It must originate through random processes, most likely through mutation and genetic recombination. Each major step would take excessively long periods of time and require many generations to accomplish, especially for major transitions. Once present, such a genetic change might be selected for differential preservation by natural selection. While such a choosing mechanism may not be random, neither is it a thinking, planning, crafting mind that guides the origination and selection process. Natural selection is not alive—it can do nothing of its own volition. Natural selection usually acts as a conservative mechanism, preserving the organism's complexity and prohibiting the major changes that evolution requires.

> Natural selection can act only on those biologic properties that already exist; it cannot create properties in order to meet adaptational needs.[2]

This brings us to an important prediction as it relates to the fossil record. If evolution were true (or even a hypothesis deserving serious study), the major transitions from one type of life to another must have been represented by actual organisms and must have taken many generations to accomplish. Since perhaps the majority of all individual organisms were intermediates, they must have at least occasionally been candidates for preservation as fossils. Of course, not all organisms were fossilized. But many were, for fossils exist today by the trillions. Thus, we would predict that some of the transitions or missing links would be documented by fossils. This would begin with the origin of extremely simple forms of life.

Actually there is no such thing as "simple" life. If it is living, it is complex. Some life forms are simpler than others, but even the simplest living organism appears to be extraordinarily engineered and complex. A general increase in complexity (more parts accomplishing more functions) would continue up the presumed evolutionary chain. For the human lineage, a list of such links in the chain would be quite lengthy, each step thought to have required many millions of years and myriads of creatures. We see examples from each representative group in the chain preserved as fossils, but are the transitional links preserved?

Missing Links

Perhaps no aspect of evolution has fascinated professionals, the general public, and schoolchildren as much as the search for "missing links." If "descent with modification" is true, the fossil record should be full of transitional forms. But it is not. Darwin himself was well aware of the problem: "Geology most assuredly does not reveal any such finely graduated organic chain; and this, perhaps, is the most obvious and gravest objection which can be urged against my theory" (Darwin, 1872, 172). Each new claim that a transitional form has been found excites a media frenzy, and the search has produced numerous mistakes and hoaxes. For instance, the famous "dino-bird" *Archaeoraptor* captured attention until it was exposed as a fraud. More recently, the small primate "Ida" was briefly celebrated as a missing link, a media darling that supposedly provided "proof" of evolution to commemorate the 200th year of Darwin's birth. (Haven't they claimed for years that evolution has already been proven?) This lasted only until more sober scientists showed Ida to be merely a variety of modern lemur.

A few candidates deserve more examination as possible "intermediate forms." Their scientific merits will be considered seriously in the following pages. Most important to note for now, however, is how the search for missing links starkly reveals evolution as a faith-based belief. Scientists should discuss evidence that has been discovered, not evidence that is missing. How could an empirical scientist know that there are supposed to be missing evolutionary links in the first place? The answer should not be "Darwin told me so." Evolutionists have great faith in the missing common ancestors needed to prove their faith. Have they found the evidence to fit that faith? That has yet to be demonstrated.

Expanding this list to include all the many necessary major transitions between animal groups—like between snails and clams, or between lizard-hipped dinosaurs and bird-hipped dinosaurs, or between the australopithecines and man—makes the chain daunting, indeed. Evolution predicts we should find at least some of these as intermediate fossils. Reason demands it.

Let us agree that in an overall sense, evolution predicts that animal fossil remains in general should reflect a simple-to-complex history of life's development. While single-cell organisms are amazingly complex, compilations of trillions of complex cells, all working together to survive (as in the "higher" mammals), are more complex yet. There should be fossils that document this overall trend. Similarly, there should be extensive diversification of both plant and animal types, from the original few types to today's abundant types. Furthermore, the developing variety of groupings should outpace extinction.

The Creation Scenario

There are numerous shades of thought among creation accounts, but the biblical account is most widely known and logical. It also presents itself as a true account of real history, one that is coherent and believable. It has stood the test of scrutiny through the years and will be the only creation account considered here.

Creation started with the Creator calling into existence "the heaven and the earth" (Genesis 1:1). From this inanimate matter, all other things were formed, including living things. Next, He created animal life. These life forms were functionally mature from the first, with fruit trees already bearing fruit, fish already swimming, and birds already flying. The biblical creation of life began with complexity. Each "kind" was created as a distinct kind, complete with all the body parts, physiology, and instincts needed for survival. There were no incipient or halfway forms, for the intelligent Creator did not need to experiment to achieve His goals. The transcendent, omnipotent God designed and called all types of life into being without having them descend from ancestors of a different type. Each kind no doubt had a robust gene pool at the beginning, enabling it to vary widely within the set limits of that kind.

Large or small, coiled or straight, ammonites have always been ammonites. This beautiful specimen was infilled by minerals after its death and burial. (Location: Bearpaw Formation, Canada. Dated at 70 million years.)

Note that the biblical word for life (Hebrew *nephesh*) is not the same as the English word for life. Life in Scripture is restricted to things with "the breath of life" (Genesis 2:7; 7:22) and things with blood in their veins, for "the life of the flesh is in the blood" (Leviticus 17:11, 14). In a biblical sense, plants are not "alive." They are certainly biologically alive and accomplish marvelous functions, but the Bible never refers to them as alive in the same sense that man and the animals are alive. Plants "flourish" for a while, then "wither" (Psalm 90:6) and fade. They were created as food for truly living things, "*nephesh* life."

The same could perhaps be said of bacteria. Again, they are biotic or biologically alive, but they possess neither blood nor breath. Thus, they are not truly alive in the biblical sense. They were created and exist to facilitate true life. Many creatures referred to as lower forms of life are also probably not living in the biblical sense. For instance, coral is classed as an animal in modern terms, but does not meet the biblical requirements for life. The same could possibly be said of insects. Although they are marvelous specimens of design and creative ingenuity, they possess neither breath (they pump air through pores in their abdomens) nor blood (their life fluids are substantially different from "higher" life forms).

The term "living" might loosely conform to the term "conscious." Only those creatures that can remember, think, and feel are truly alive. Animals can do all these things, although to a lesser extent than humans can. We all know dogs that have personality and emotion. They are certainly alive, but unlike humans, they are not capable of forethought and reason.

The speculation that plants and "lower" animals do not possess biblical life may not seem important, but it answers a question that some have raised. The Bible's creation account implies that in the very beginning—before Adam sinned and incurred "the wages of sin" (namely, death; Romans 6:23)—there was no death of conscious life. But when humans eat plants, it requires the biological "death" of both the plants and certain digestive bacteria. When we perform many everyday functions, such as walking, certain insects may be casualties. Defining life as the Bible does answers the seemingly problematic question of whether death was part of the original creation.

A third "creation" was accomplished when God "created man in his own image...male and female created he them" (Genesis 1:27). Many traits are no doubt implied by the "image of God," but suffice it to say that man's spiritual nature is not shared by the animals, and it adequately reflects God's image.

In the beginning, different types were often dependent upon each other through mutually beneficial symbiotic relationships. Each was a necessary part of an ecosystem. For instance, mammals possessed the necessary internal bacteria to accomplish digestion.

For those kinds that employed sexual reproduction, both male and female had to be present. Mankind, specifically Adam and Eve, were given stewardship over creation, to care for it wisely for man's good and the Creator's glory. Everything that was needed was provided, fully accomplishing the Creator's purposes. There was nothing of Darwinian struggle and death in the world God created as "very good" (Genesis 1:31).

The creation account is augmented by further events that continued to shape the world. According to Scripture, the perfect creation was interrupted by the entrance of imperfection. A universal introduction of death, decay, and deterioration followed Adam's rebellious rejection of the Creator's authority, and soon all things suffered under that penalty (Genesis 3: 14-19). Our present "fallen" world, unlike God's original "very good" creation, is full of Darwin's "war of nature," disease, and death.

Scripture likewise tells of a worldwide deluge (Genesis 6-9). This hydrodynamic cataclysm was described as nothing less than a global tectonic restructuring of at least the upper layers of the entire earth. It left in its wake mountain chains, volcanic arcs, and separated continents. All living things that inhabited dry land succumbed to its waters, except those on board the Ark. Most important for this discussion, the Genesis Flood produced vast deposits of water-laid sediment that was full of dead plants and animals, now hardened into sedimentary rock and fossils. No place on the planet escaped the Flood's terror. Everywhere we look, in every rock or fossil deposit we study, we encounter flooded terrain.

Some fossils no doubt date from before or after the Flood, but the vast majority are the direct results of sequential flood processes. First, sea-bottom dwellers were transported and buried in marine sediments. As the Flood encroached farther and farther onto land, sediments would be increasingly terrestrial, entombing land animals along with the abundant marine creatures. All were originally living at the same time, and then died and were buried in successive Flood activities.

Obviously, one's worldview intertwines with one's science in creation thinking, but the same is true of evolutionists. Remember, in origins discussions, unobserved events are under consideration, and both sides must fill in the gaps with a "story." Neither side witnessed the origin of any separate and distinct plant or animal type, nor its preservation as a fossil, and must therefore speculate what brought it about.

The evolutionist's story speculates about highly improbable events, such as random but beneficial mutations that produced new genes with increased information content. The theory must continually be expanded and amplified to accomplish evolution's marvels, and yet such unthinking and random natural forces appear to be woefully inadequate for the task. Sole reliance on natural laws to produce an ordered universe reflects a worldview commitment to materialism or naturalism—in other words, a faith position.

Creationists presuppose a God powerful enough and intelligent enough to accomplish creation with all its intricacies. They do not attempt to prove His existence, but assume it and allow this assumption set to inform their science. The Creator presents Himself in His written account as completely knowledgeable and fully trustworthy, incapable of misrepresenting truth. The Bible is not hard to understand, although at times it is far beyond our experience and thus hard to comprehend. Our incomplete comprehension must always be willing to be modified as new information becomes available, but whenever a testable statement is made in His written account, we find that His assertions match reality. Believing the creation account of Scripture may require faith, but it is a reasonable faith—unlike the incredible faith of the evolutionist, for modern-day processes are seemingly incapable of accomplishing the ultimate origins of things.

Let us summarize the predictions of each model as they relate to the fossil record.

Evolution's Predictions

- All things originated and developed through time, chance, and natural processes based on the properties of matter.
- Basic types of organisms developed in stages from previously existing types.
- Basic types proliferated over time.
- Complexity increased over time.
- Living things were buried and fossilized by processes possible today.

Creation's Predictions

- All basic types originated abruptly through supernatural processes.
- Organisms diversified within limits from original created kinds.
- Each kind experienced either stasis or extinction over time.
- Complexity was present from the start.
- Burial and fossilization occurred through extreme processes that are unlikely today.

Fossil flowers may be even more supportive of creation than fossil animals. They are even less supportive of theistic evolution and old-earth creation. Much evidence can be seen for variety, but not for the origin of new basic types. In evolution, atheistic or theistic, flowering plants supposedly did not appear until long after the land animals. But according to Scripture, flowering plants were created on Day Three, while land creatures were created on Day Six. Any system that includes great ages is incompatible with Scripture.

Chapter 4 Time

Embedded within the creation/evolution issue is the matter of time. While in principle creation could have taken place over any length of time, evolution requires deep time—sufficient time to accumulate enough unlikely beneficial mutations to accomplish the necessary transitions. There must also have been enough time for all harmful mutants to have been eliminated (i.e., to have died and gone extinct) through natural selection.

Time especially enters the discussion when considering fossils. Evolution charts evolutionary change according to the time represented on what has become known as the geologic column (see the chart on page 28). Each of its eons-long epochs, periods, and eras is referenced to the particular life forms that are believed to have lived during that time, as represented by their fossils. Fossils found lower on the column are thought to have lived earlier in earth's history than those found higher.

As evolutionists present it, the bottom of the column represents the early earth, indicating at first an absence of life, and then the generation of life from non-life. Above that, the evolutionary story continues, with the sporadic appearance of fossils of single-cell life. Then, proceeding upward, multi-cell invertebrates flourished, followed by fish, amphibians, reptiles, birds and mammals, and finally man at the column's top. The fossils in the column are thus displayed in an evolutionary progression.

As we shall see in the discussions to follow, the fossil record is very different from this simplistic chart. Rather, the chart—which appears in virtually every book and textbook on this or related subjects—is really a statement of evolutionary dogma. Without a doubt there is order to the fossils, but this chart does not tell the whole story. Does that order reflect stages in the Darwinian struggle of life and death, producing "new and improved" life forms over millions of years, or does it represent stages of the great Flood? The true record of the fossils is better represented by the right-hand column of the chart on page 28. Almost all fossils are marine invertebrates, mostly animals with hard outer shells, like clams. Invertebrates (including clams) live today in abundance and have done so throughout every epoch of earth's history, but certain individual varieties recognizably lived during specific periods. Thus, they merit being called "index fossils" and are used to identify the particular time in which they lived.

Uniformity versus Catastrophe

There is more to understanding the geologic column than merely the length of time it depicts. Rather, it represents an attitude toward time and what may occur in it. Remember that the history it charts is long ago, supposedly before any human was present to observe what sorts of creatures were alive. The evolutionary worldview thus necessarily stems, not from scientific observation, but from beliefs about the remote past—in particular, the idea or assumption of "uniformity."

Uniformity (or uniformitarianism) in general assumes that what occurred in the geologic past was not much different from events that are possible in the present, especially with respect to the laws of nature—thus restricting thoughts of past processes to similar process rates, scales, and intensities observed today. By extrapolating present processes into the past, one can surmise certain things about past events. Obviously, the extrapolation is only as good as the assumptions on which it is based. What if the assumptions are wrong?

A good example of applied uniformity can be seen in Grand Canyon. Imagine two scientists, a creationist and an evolutionist, standing on the rim of the canyon, trying to decipher its history. They ask, "How did the canyon form and how long did it take?" Certainly, the canyon has been eroded, but neither scientist actually witnessed the erosive event. Nor can anyone else provide a reliable eyewitness account. How could either one possibly know for certain what was involved in this unobserved past episode?

Both observe the same facts about the canyon. At eighteen miles wide at the widest point, about one mile deep and 150 miles long, the canyon dwarfs both view-

Geologic Timescale with Dominant Fossils

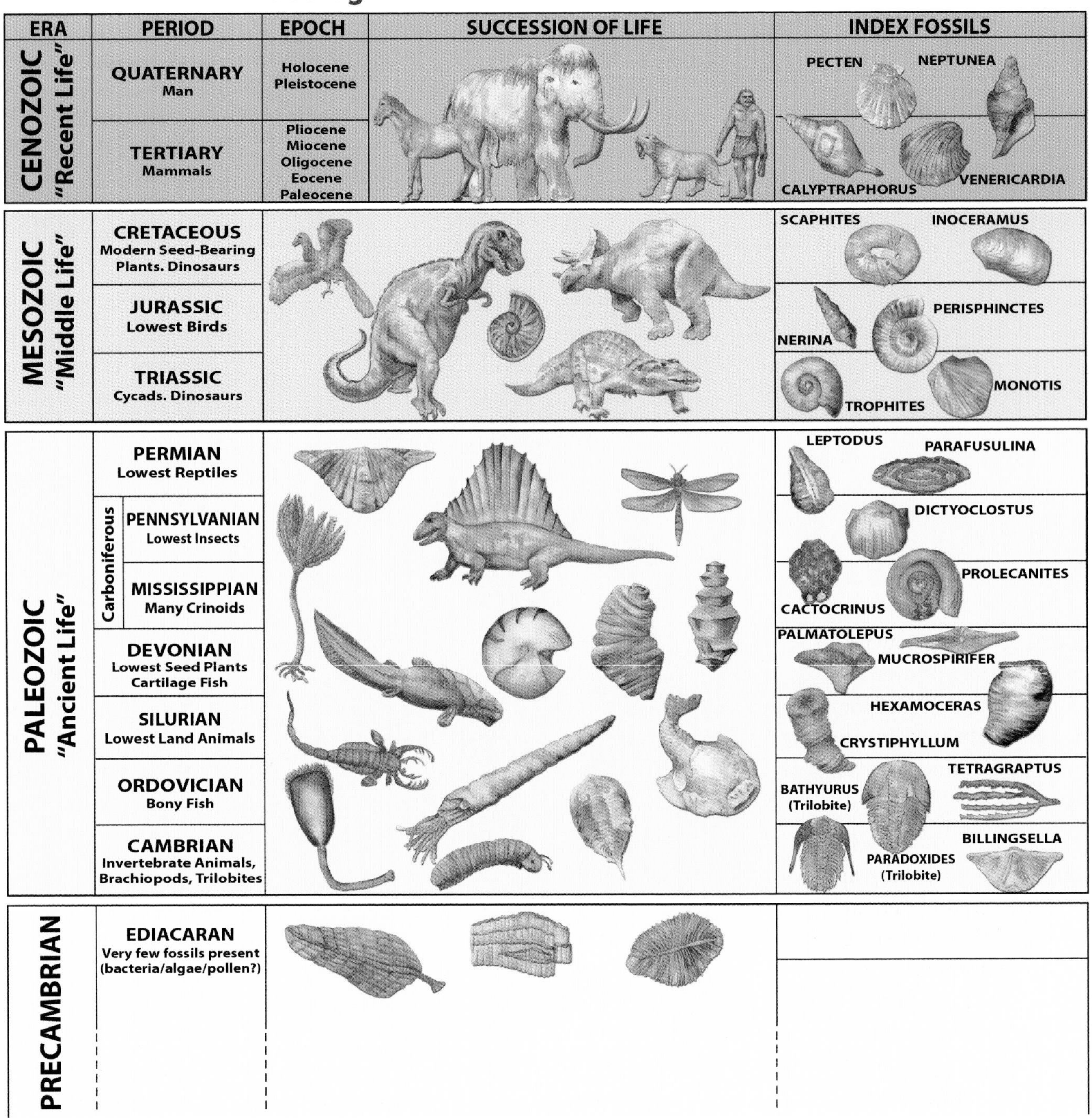

The geologic column as standardized by the Geologic Society of America in 2009. The wording and ages for Precambrian Time have not reached a consensus. Remember that although this is presented as a column of fossils, it is better understood as a column of proposed evolutionary development.

ers. They measure the present width of the Colorado River and its power to erode. They observe how the process varies over the years, and obtain a record of historic river floods. In flood stage, much more erosion takes place than during normal years, when perhaps more silt is deposited than taken away. Based on their observations, an average is calculated. Even though the two scientists differ in their worldviews, they agree 100 percent on present processes and process rates. But not much has happened to the canyon in observed recent times, and they are concerned with the unobserved, long ago past.

To determine the canyon's origin, the evolutionist starts with the assumption of uniformity. The river, which currently seems to deepen and widen the canyon ever so slowly, is assumed to have done likewise throughout assumed past ages, operating at a similar average rate of erosion that is extrapolated into the past. He does careful and accurate work, and concludes that the river-carved canyon is of great age, having been excavated by

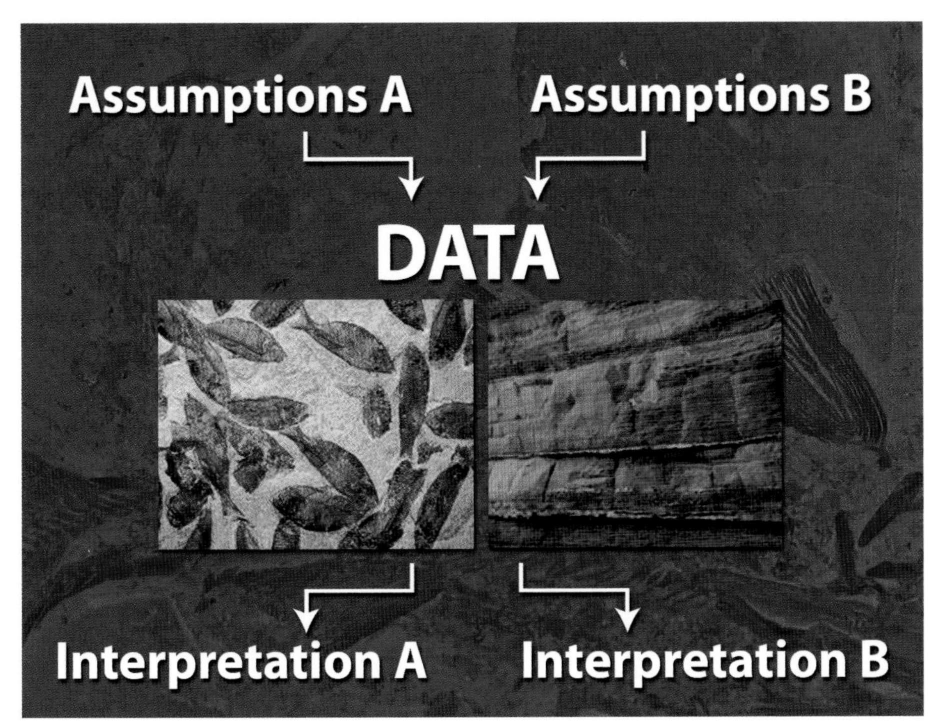

The study of data must inevitably begin with a set of assumptions, which dictate which data are deemed important, which measurements are chosen, and which experiments are run. The interpretations follow, but the assumptions, or worldview, held at the start are easily the most important part of the interpretation process.

an immeasurably slow process that accomplished great amounts of work over long periods of time. It would help if he had a reliable eyewitness to consult regarding his conclusions, but no one was present when the canyon was formed. He has reasoned with elegance and consistency, but is he right?

The creationist matches his colleague regarding careful observations. But he employs a different assumption set, based on an alternate worldview, and comes to a radically different interpretation. Rejecting the assumption of uniformity, he employs the principle of catastrophe, which holds that processes may have operated in the past at much different rates than those that are observed or are potentially possible in the present. He points out that different process rates were evidently needed, since present processes could seemingly never accomplish the work that has been done. Maybe different processes or process rates were involved.

Perhaps he is certain that God created all things in the past using creative processes that are not now in operation. Furthermore, he may have been convinced by Scripture that the great Flood of Noah's day subsequently inundated the entire world (including Arizona, the home state of Grand Canyon). He knows that cultures worldwide, all of whom descended from the survivors of that Flood, contain a "memory" of that event in their folklore. Scripture does not once mention Grand Canyon, but from what we know about big floods, the great Flood would have left traces that are still visible today. Indeed, that Flood and its aftereffects would have restructured the entire planet. The geologic processes involved may be recognizable to modern geologists (erosion, deposition, fossilization, etc.), but such a high-energy event would have sculptured earth's surface using processes operating at far greater rates than those possible today. Thus, the canyon might not be so old.

It either took just a little water and a long time to produce the eroded canyon, or it took a lot of water and a short time. These differing conclusions are not simply a matter of faith versus science, as evolutionists often claim, for both sides exercise faith regarding the unseen processes that acted in the past. Nor is it a matter of the quality of scientific observation, for both sides make essentially the same observations. Rather, two different ways of thinking about the past are on display. Neither is capable of absolute proof, for no one can travel back in time and actually observed what happened. The only thing that can be done is to determine which set of assumptions makes "predictions" that better fit the evidence, and is thus is more likely to be correct.

Note that creation scientists/Flood geologists do not simply say "God did it" to explain what they observe. They refer to the same natural processes uniformitarians do. However, they point to the evidence that shows that these processes operated catastrophically.

There are other observations that have a bearing on this

issue. For instance, measurements have determined that very little, if any, erosion takes place along a slow-moving stream bed under normal circumstances. When mud and silt cover the river bottom, the water and suspended load scarcely ever interact with the underlying rock. However, when a mighty rain or a river flood causes great volumes of water to flow at an accelerated rate, the abrasive grit and rocks carried by the water scour the mud off the river bed, allowing erosion of the underlying rock to occur. Over a long time there could be many floods, so significant work could be done. But not even large floods compare to the deluge mentioned in Genesis. Those flood processes would have acted at rates, scales, and intensities not possible today, and its effects would dwarf the cumulative effects of many smaller floods.

Modern scientists did not observe that great Flood, and it is impossible to fully grasp the impact it would have had on a local system. Much research into major flooding has been conducted by today's engineers, and much more continues, but from Scripture we can certainly glean that its effects would have exceeded our imaginations.

Remember that the geologic column is primarily a column of time, with the older layers and fossils situated beneath the more recent ones. Often, fossils are referred to as "appearing" in the record. This refers to the discovery of the stratigraphically lowest, or "oldest," fossil of that type. According to evolutionary thinking, that type of creature probably did not exist during earlier periods, which are lower down the column than the layer in which it was found. Or if it did exist earlier, it left no fossil trace. Its appearance is followed by the length of time, or range, its fossils continue to be found in the overlying layers of the geologic column. After appearing in the record, all fossil types either survived into the present or fell prey to extinction.

How do you date a fossil in order to place it in its appropriate timeframe in the geologic column? Many people have the misconception that fossil age can be determined by radioisotope dating. These techniques, however, are applicable only to rocks that were once in a hot, molten condition, such as lava rocks or granite, which seldom contain organic remains. Sedimentary rock, which contains nearly all fossils, cannot normally be dated by this kind of technique, for it consists of

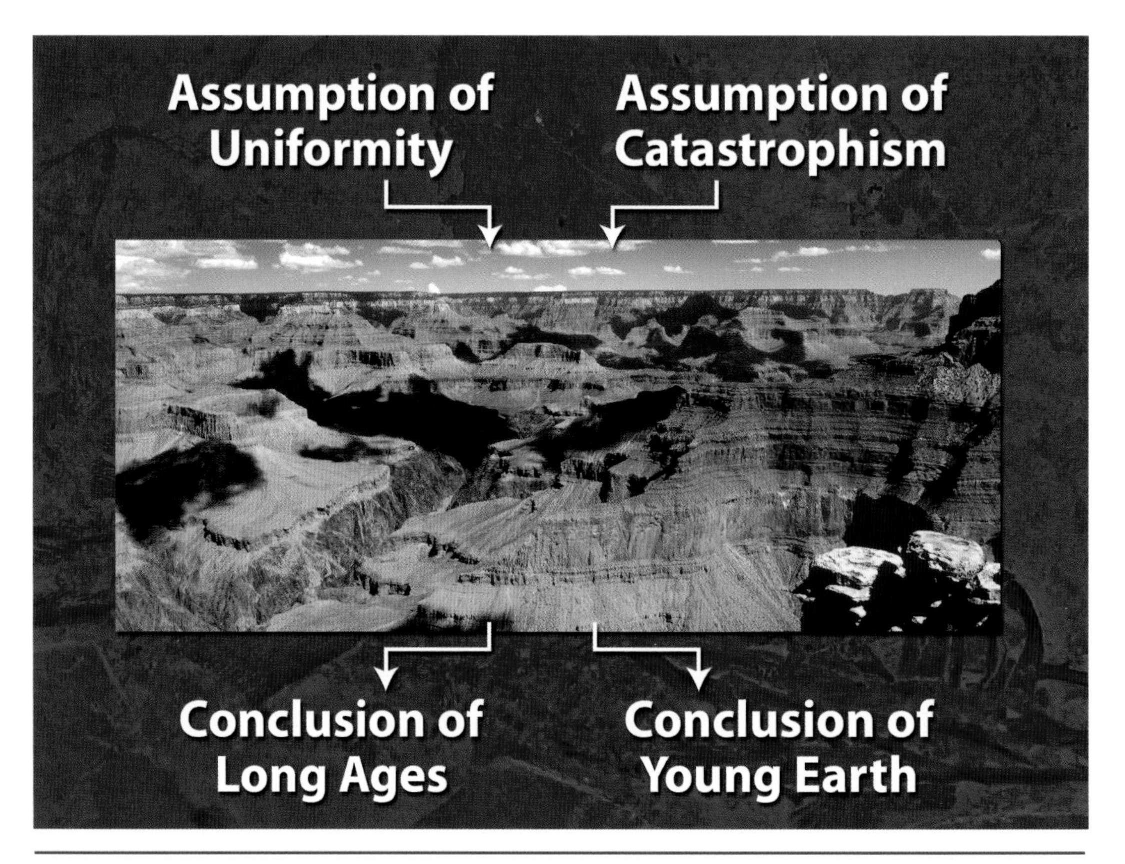

Grand Canyon—the grandest of canyons. Here we see multiple layers of fossil-bearing sedimentary rock dissected by the Colorado River. Interpretations range from millions of years of formation time to a short time during the great Flood of Noah's day. Both conclusions use the same data. The different interpretations stem from different assumptions that are applied at the start.

eroded particles or chemicals dissolved from previously existing rocks. Eroded sediments are transported and redeposited, so the rock materials they contain existed before the present layer in which they are located.

Carbon-14 dating is sometimes used to date organic remains, but the organic material from the original creature or plant has likely been removed or contaminated by the subsequent fossilization processes. Thus, its use is also not suitable for dating. Nevertheless, although few fossils can be accurately dated, nearby igneous rocks can be dated (theoretically, at least) and their dates applied to adjacent fossil-bearing strata.

But can even igneous rocks be accurately dated? While a full discussion of radioisotope dating will not be attempted here,[1] it must be noted that this dating method is usually applied to rocks that are thought to be of extreme age—so old that no one could truly know their age. Unobserved events and times are being measured, so certain assumptions must be made about the unobserved past. Those assumptions must be correct if we are to derive accurate ages. The assumptions normally used in radioisotope dating will be recognized as conforming to the principle of uniformity.

In general, radioisotope dating efforts proceed by recognizing that some atoms commonly found in igneous rocks are radioactive and unstable (e.g., uranium). They spontaneously decay into smaller, stable atoms (e.g., lead), giving off radiation in the process. We can accurately measure the present amount of both "parent" and "daughter" isotopes, as well as the present rate of decay of one into the other, and this allows us to calculate the apparent age of the rock. Let's list the assumptions inherent in this calculation.

In radioisotope dating, the assumptions are:

1. The rate of decay has been constant throughout the entire time the rock has existed.
2. The quantity of parent or daughter atoms has not

Metoposaur skull, about 20 inches long. This great amphibian would have been eight feet long. Its fossil was found in the west Texas Triassic Red Beds.

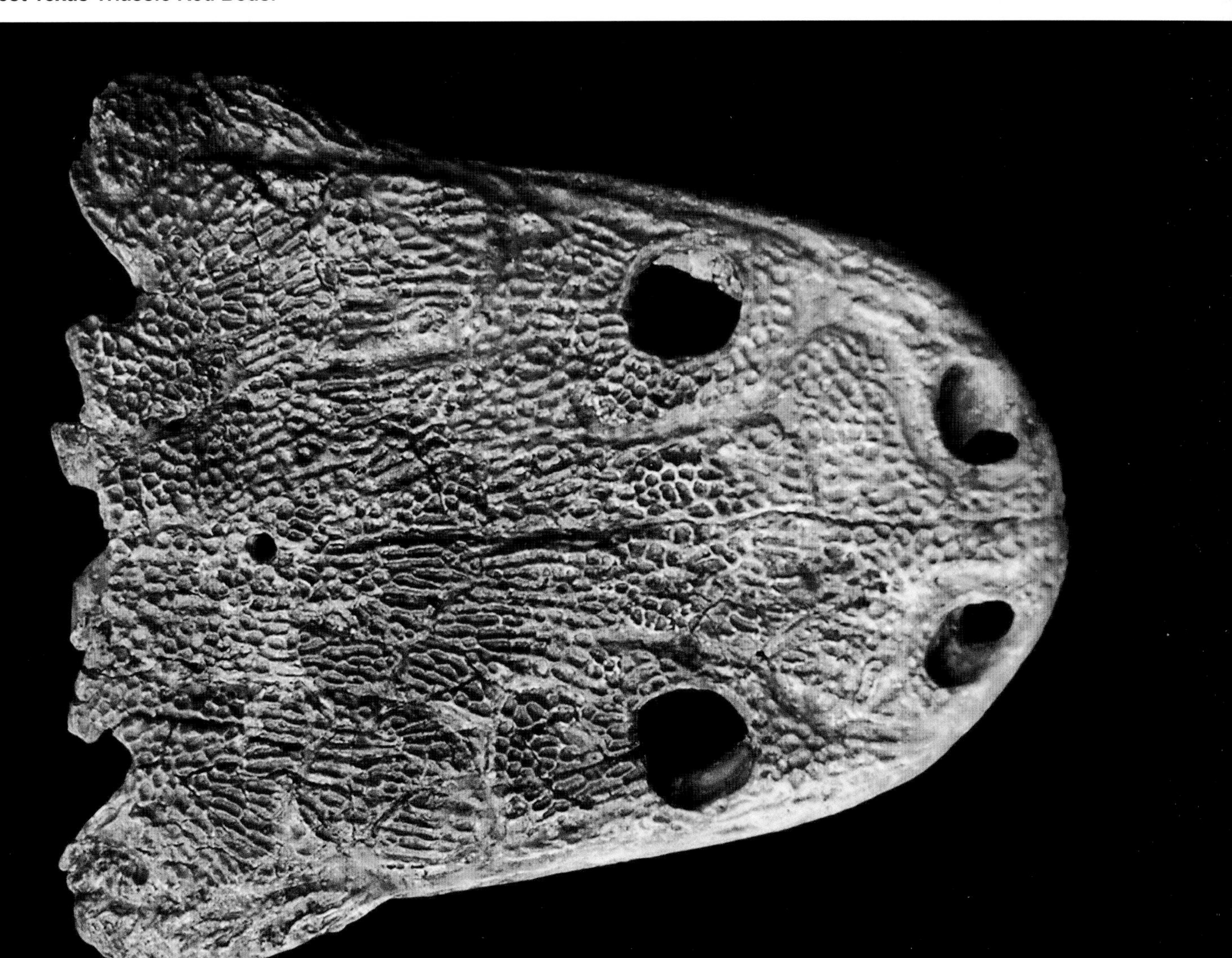

been altered by outside contamination.

3. The original quantity of parent and daughter is known.

All three of these assumptions have often been shown to be in error.

1. Evidence has been discovered that shows decay rates to have been quite different in the past. To assume otherwise is to ignore much data and adopt the questionable assumption of uniformity.
2. Many ways are known through which both parent and daughter atoms can be either added to or leached out of the system. Since most have to do with water action, consider what effect the Flood might have had. No place on earth would have been immune to its effects.
3. Igneous rocks that were observed to form in historic times have isotope concentrations that are unexpected and that are more indicative of rock that has been around for "a long time." When rocks are first formed, they often look "old," if the above assumptions are applied.

To summarize, all of the assumptions used in radioisotope dating are questionable and commonly give erroneous, inconsistent, and sometimes bizarre dates. It is not an infallible method by which to date rocks.

Ignoring the unreliability of radioisotope dating, evolutionists have observed the strata, arranged them into a stratigraphic column, assigned ages to them, and constructed the unobserved geologic timescale. The chronology can be refuted (as it is in other treatments) from without and critiqued for internal consistency. But for this discussion of the fossils and their relative consistency with evolution, let us assume it to be correct as presented. Let's see if evolutionary thinking is internally consistent.

The Cenozoic Era, presented as comprising the past 65 million years, is thought to represent "recent" life, including the rise to dominance of mammals and, in its later stages, man. Before it came the Mesozoic Era, or the time of "middle" life, representing a time period from 65 to 230 million years ago and including the dinosaur "age." Still earlier came the Paleozoic Era, hosting "early" life, from 230 to 570 million years ago (the exact numbers are still debated among evolutionists). This period contained abundant marine creatures and ancient land creatures. Earlier yet came the Hadaean, Archaean, and Proterozoic Eras, during which presumably life originated from non-life.

***Dickinsonia costata.* The Ediacara biota is difficult to place within any known group. This tiny living thing was deposited in the pre-Flood world or during the very earliest Flood events. It may provide evidence of a created habitat and life assemblage that are completely unlike any found today.**

Evolutionists have always been somewhat chagrined by the sudden appearance of so many animal phyla, or basic body plans, in the Cambrian period of the Paleozoic Era. In lower strata, fossils are much more simple and rare. It has been contended that earth experienced a sudden proliferation of complex life forms in an event termed the Cambrian explosion of life. Recognized as a serious problem for evolutionary gradualism, it spawned an intense search for the ancestors of Cambrian life.

Uniformitarian thinking predicts that fossils would be buried near to where they had lived by the local processes observable today. They would only rarely be mixed with fossils from a different environment. Since many carcasses would have been exposed to scavengers, bacterial decay, and oxidation agents, the required conditions for fossilization would not often have been satisfied. The bodies of individual animals might be scattered, as would communities of larger populations of animals, rather than be concentrated in any given area. Since great time would have passed during the different stages of evolutionary development, the fossils characteristic of one time period could scarcely have been represented in the sediments of another, far removed time period. Some fossils and depositional events would be constant throughout time, but in general, there should be regularity that is consistent with evolutionary and modern processes.

Predictions based on creation/catastrophe thinking are quite different. The Genesis Flood would have annihilated nearly every living thing on land, and many things living in the sea. Sedimentation would have been extremely rapid, favoring the considerable preservation of dead remains. Living organisms may have been buried in mass graveyards, which would contain remains from differing habitats. Usually a carcass would have been entombed with others from its local environment, but plants and animals from different environments may have been buried together. As the "marine-based" Flood continued to cover the entire earth, sediments and plants and animals would exhibit more and more terrestrial traits, although they would remain dominantly marine throughout. The expected evolutionary order would not necessarily be followed. The fossils would instead be in the order of flood inundation and deposition. Life forms would show complexity and design at every level, supportive of a past creation event rather than long ages of progressive evolution.

Having made these two sets of predictions, we can now visit the real fossil record and determine which set better reflects reality, and is thus more likely correct. In the following chapters we will attempt to shed some light on that question. But first, let's list the types of fossils.

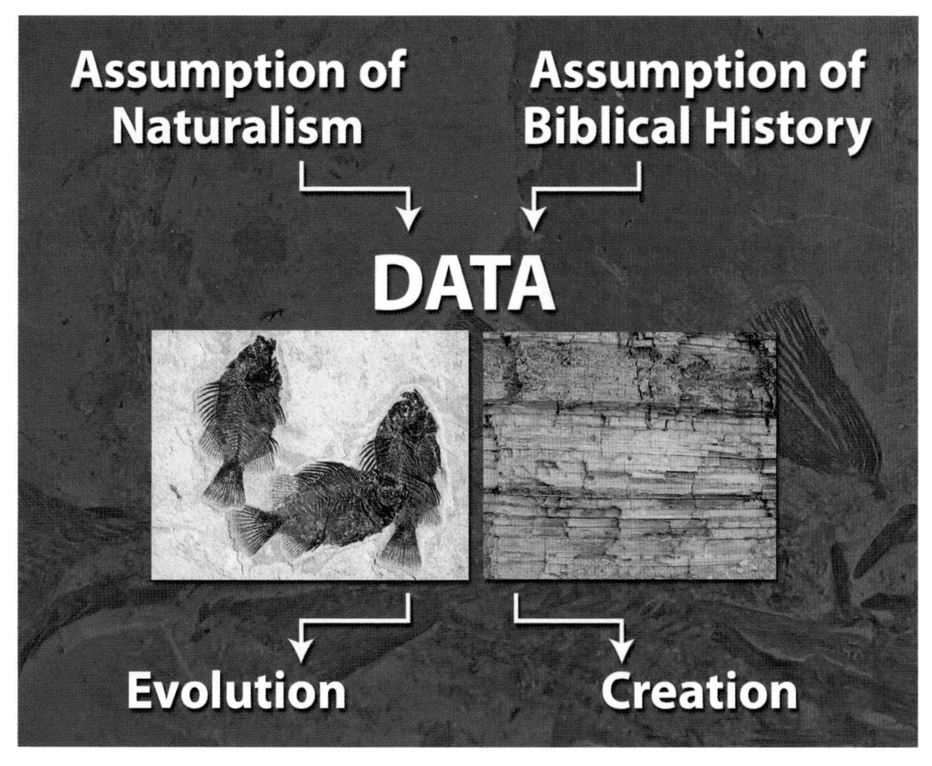

By assuming naturalism and uniformity, one can study the data and arrive at an evolutionary interpretation. But this is not the only possible assumption set. If one adopts biblical history as true, especially creation and the Flood, the data can be interpreted within a creation context. Comparing the two methods and conclusions shows the creation/Flood worldview to be more scientifically viable.

Chapter 5 The Nature of the Fossil Record

This book will concern itself mainly with the message of the fossils—i.e., the fossil record. Events regarding fossil deposition deal primarily with the long-ago past. The word "record" normally implies a written account composed by a knowledgeable eyewitness, but we are not afforded such an account if biblical history is disallowed, as demanded by the naturalist. Nevertheless, we will try to discern what the fossils can relate to us, and therefore will refer to it as a "record." The appendix will provide more specific information about individual vertebrate fossils that have been considered to be "transitional forms," but the bulk of the book will concern itself with larger trends, such as whether the record is more compatible with and supportive of the evolutionary view of history or the creation view of history. Does it speak of normal, uniformitarian events, or of one-of-a-kind catastrophic events?

We must first define what a fossil is. There are many different types of fossils. Every trace of the physical existence of a once-living plant or animal that is older than recent remains can be deemed a fossil. How can a living thing be preserved once it no longer lives?

Consider what normally happens when a plant or animal dies. Unless fortuitous conditions exist, the remains of a living thing will begin to decay soon after death. Whether a small rodent in the forest, a jackrabbit in the desert, a fish in the ocean, or a fern in a swamp, the remains will soon be attacked by scavengers, undergo disintegration through oxidation, or be decomposed by bacteria.

Just a few decades ago, millions of buffalo roamed the American prairie. Today only a few thousand remain under government protection. When the great herds were rampantly slaughtered, their carcasses lay piled on the ground. Where are their fossils? Except for a few skulls and teeth that were buried by small floods, their remains quickly disintegrated. In nearly every case, the fossil of a plant or animal must have undergone rapid burial in order for it to have resisted full destruction.

An oft-repeated series of textbook illustrations shows a hypothetical animal dying alongside a stream. Before nature's degradative influences have taken hold, the stream overflows, burying the carcass in mud and protecting it from ruin. Over the years, mud accumulates around the remains. Eventually, the entire region subsides, allowing even greater thicknesses of lake bottom or ocean bottom mud to blanket the area, mineralizing the bones and consolidating the mud into rock. When the region rises again, erosion exposes the now-fossilized remains.

This scenario would, no doubt, be applicable in rare cases, but it ignores significant advances in sedimentation theory that have been made in recent decades. Geologists now recognize that most rock units resulted

Jumbled together in a bone bed near Agate Springs, Nebraska, are remains from an amazing assortment of mammals that date from the final stages of the great Flood of Noah's day. No such deposit is forming today. The present is not the key to the past, as uniformitarian thinking insists.

from widespread, high-intensity processes that can accomplish in minutes the work that had traditionally been attributed to slow and gradual processes. Turbidites, underwater gravity flows, comet impact, megavolcanics, hypercanes, and other major catastrophes, both regional and local, are often called upon to account for the deposits.

Not only do the rocks speak of such major geologic episodes, but the fossils themselves seem to demand rapid, laterally extensive, dynamic processes. They are frequently found in vast quantities and with exquisite preservation, quite unlike the products of slow and gradual transformation. This is true for all types of fossilized remains.

Types of Fossils

All fossil categories can be interpreted as resulting from either catastrophic or uniformitarian processes, but the question arises: Which model provides the better explanation in the majority of cases and is therefore more likely correct? Which idea about history can account for the varied fossil types?

Unaltered Hard Parts: Often, the original organic material remains after fossilization. Most fossils are of marine shellfish, and their unaltered shells are commonly found, usually consisting of calcitic (calcium carbonate), phosphatic, or silicious shells. This category also includes bone, coral, and wood.

Comprehending the processes involved in strata deposition and fossil burial and preservation is a challenge. The conditions and scenarios required are very different from corresponding processes today. The trail of dinosaur tracks, found in central Texas within the laterally extensive Glen Rose Limestone, contains over one hundred prints. Deposition demands regional flooding, but the tracks must have been made in soft mud, before the sediments could harden. Preservation speaks of infilling before erosion. The fossilized worm burrows in the upper right image imply extensive bioturbation while recently deposited muds were still soft. These ephemeral features can be found in numerous locations and strata. The marine invertebrate in the lower right, a gastropod, reminds us that the overwhelming majority of fossils found on the continents were not of land animals, but of ocean creatures, primarily shellfish. Taken together, fossils require major flooding, rapid burial, and quick hardening of sediment.

Altered Hard Parts: Once buried, the fossil may undergo alteration of its hard parts while its soft parts decay. Organic material may be replaced or encapsulated by minerals that have been dissolved in the percolating groundwater, a process termed permineralization. If the inserted mineral is silica, the process is called petrifaction.

Soft Parts: More rarely, the soft parts of an organism are preserved through various processes. When an animal or plant is suddenly frozen and isolated from bacteria, its soft, original, organic makeup may remain. Similarly, natural mummies are formed by dry air and heat. Better known are the fossilized insects that were entombed in amber, or tree sap.

Cast or Mold: In another common fossil type, the organism decays and leaves an impression in the surrounding sediment that preserves the original shape. If a mineral fills the resulting cavity, it is called a cast.

Carbonization: When heated, buried plants expel their gasses, leaving behind mostly carbon. This results in the formation of coal, in which the parent plant can often still be recognized. Animal bodies are sometimes flattened by weight, with only a thin carbon film remaining.

Coprolites and Gastroliths: Fossilized fecal material (coprolites) or fossilized stomach contents are sometimes found. The undigested food can often provide evidence of the animal's eating habits. Some animals swallow rocks to aid in digestion, and thus gastroliths are also found, rounded smooth by digestive acids.

Ephemeral Markings: Animal tracks and trails are eas-

(Below Left) The Florissant Fossil Beds sport numerous tiny insects trapped within finely laminated sediments. A volcanic source for much of the sediment has been determined, while flowing water and mud produced multiple tiny layers. (Location: Colorado. Dated as Eocene.)

(Below Right) A coal seam near Price, Utah, is sandwiched between marine strata. Coal is the metamorphosed remains of pre-Flood continental plant material. The knife-edge contact above and below the coal belies any long-age interpretation of the formation of either the coal or the marine strata. (Dated as Upper Cretaceous.)

ily eroded and normally exist for only a short time. Unless quickly buried and filled with sediment, they will not last. The same can be said for worm burrows and raindrop impressions.

Widespread Fossil-Bearing Strata

It is helpful to think in three dimensions about geologic layers and the fossils they contain. Generally, sedimentary rocks are deposited as laterally extensive, flat-lying blankets of mud, often containing the organic remains of plants and animals. Sometimes these blankets are hundreds of feet thick and hundreds of thousands of square miles in area. Under the right conditions, they harden into sedimentary rocks and the organic remains become fossils. Unless disturbed by subsequent forces or dissected by erosion, they will continue for some distance. Each bed is rather unique and can often be traced laterally by simply observing its particular characteristics and fossil content.

Early geologists recognized that the water-driven geologic process that deposited each layer had carried recognizable sediment and organic content, likely from the same source area. Thus, fossils came to be used as a diagnostic feature of a given bed. Knowing the fossil content of a bed in one area might help identify that bed in another location, even if it has been separated by erosion.

The continuity of rock layers cannot be directly observed because they are typically located deep underground. The digging of oil wells, however, has proven to be very useful in this regard. The drill cuttings of the layers, complete with their unique chemical and fossil content, are washed to the surface. This allows the strata to be identified.

Nearly all sedimentary rock was deposited as sediments in moving water or air. The majority of fossils have a marine origin, but the fossil content becomes more land-derived higher up in the geologic column. For instance, the fossils of land-dwelling dinosaurs almost always overlie remains from marine environments when both are found in the same geographic location.

Typically, fossils of any habitat are found in geologic units that are widespread over vast areas, quite unlike sediments of today. The rock strata often bear evidence that regionally extensive catastrophic processes were involved in their deposition, with rates, scales, and intensities unlike those operating today. Something radically different shaped the unobserved past.

Consider the Jurassic-dated Morrison Formation and its thick Brushy Basin Member, which contains abundant dinosaur fossils mixed with petrified wood, clam and snail fossils, and others, including quite a few mammal fossils.[1] Creation geologists Bill Hoesch and Steve Austin have determined that all are found in a mixture of altered volcanic clay, sand, and chert,[2] which was brought in by flowing mud from a distant source. The tremendous volume of the clay—more than 4,000 cubic miles—has given rise to the concept of explosive mega-volcanoes. These involved the eruption of enormous quantities of ash, and flowing mud that re-

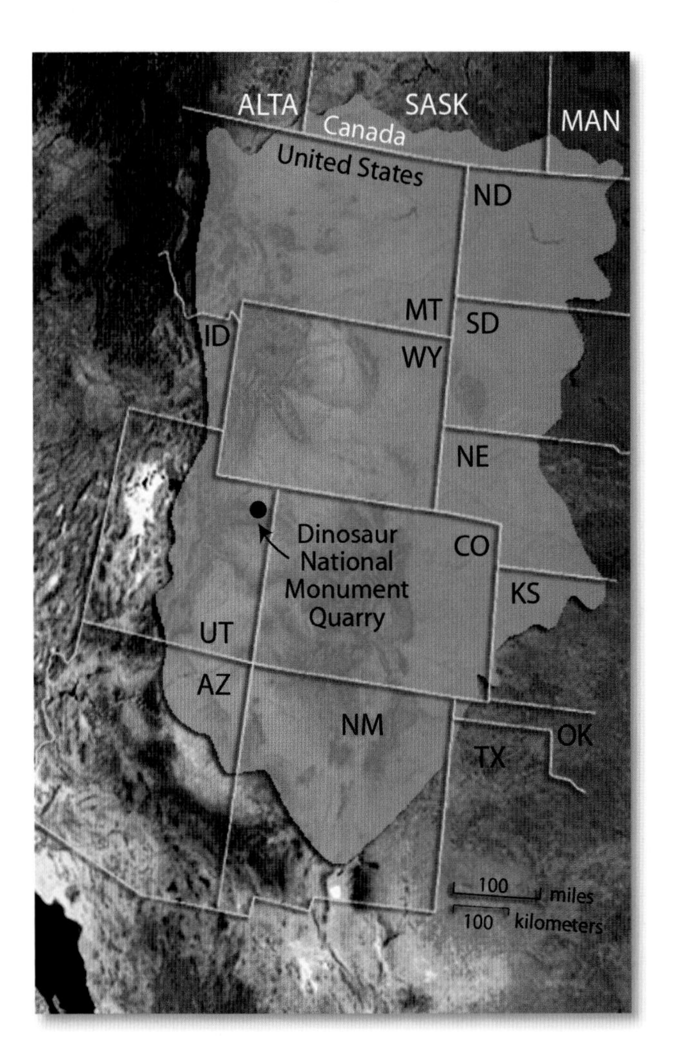

The Morrison Formation, which covers much of the American West, contains abundant dinosaur fossils. Its source has been interpreted as mudflows from a volcanic event, yet it also contains many marine fossils, such as clams. Its contents hardly comprise a complete ecosystem. What would the numerous dinosaurs eat? Some consider the extensive layers to be the result of a migrating stream, with many huge dinosaurs caught in local floods. Is there a better interpretation—such as widespread, catastrophic flooding that deposited many fossils on the continents, all entombed in a mixed environment? Could it have resulted from the great Flood of Noah's day?

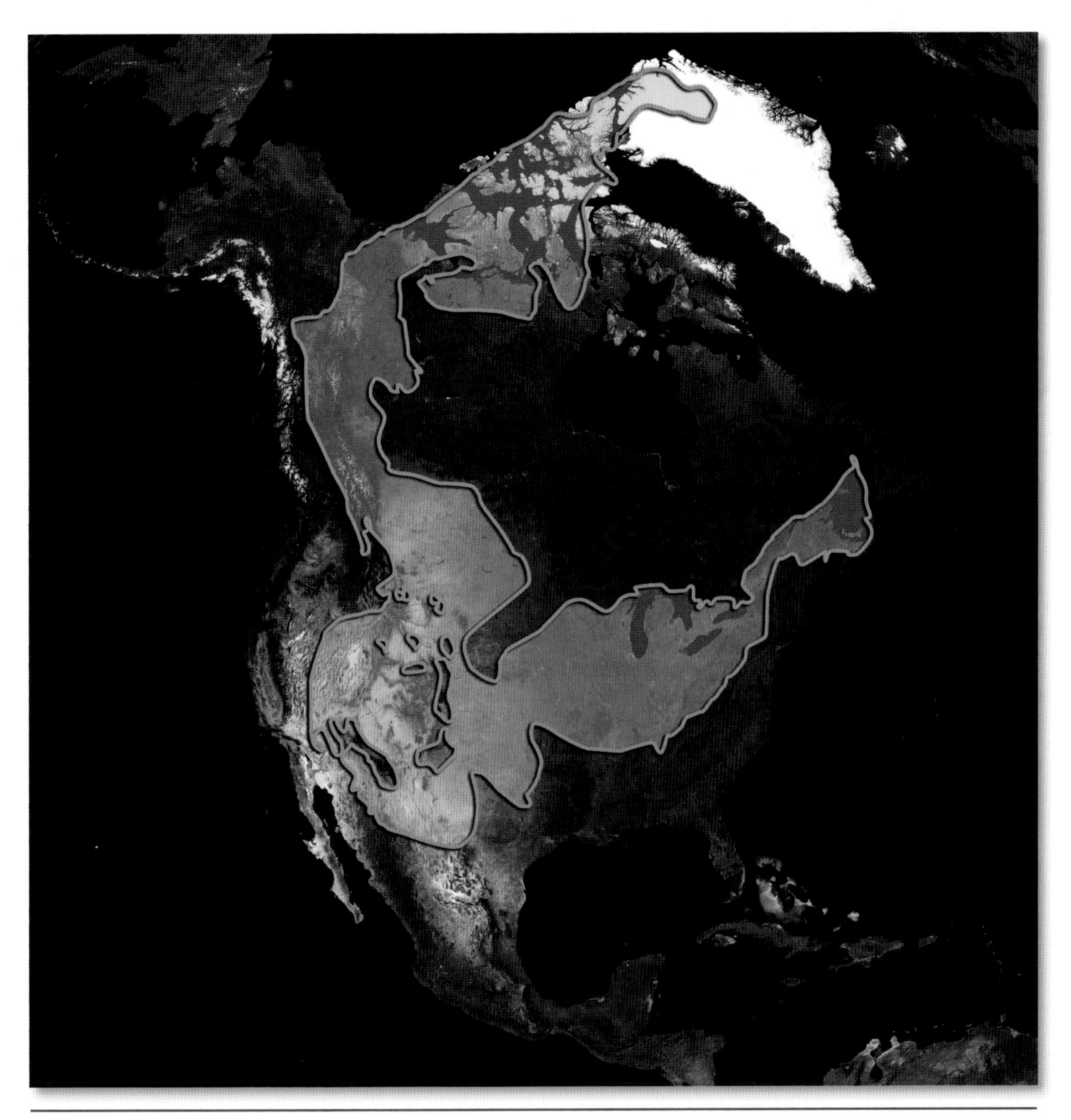

The Tapeats Sandstone is the lowest and oldest horizontally bedded strata in Grand Canyon. Geologists often interpret it as having resulted from dynamic underwater mudflows. Yet its lateral extent is continental or hemispherical in scope. Something different was going on in the past than what evolutionists assume.

sembled enormous floods, dwarfing modern volcanic events. A map of the lateral extent of the Morrison Formation shows that it covers an area of over 700,000 square miles from New Mexico to Canada. No uniformitarian process comparable to today's episodes caused this deposit.

The Morrison layer once blanketed an even greater area. It is likely that it covered the American West and overlay the familiar layers of Grand Canyon before being removed by erosion.

Notice Grand Canyon's stratigraphically lower Cambrian rocks. The Tapeats Sandstone, the Bright Angel Shale, and the Muav Limestone and their correlating beds throughout the northern hemisphere comprise the Sauk megasequence. It too must have been deposited by unthinkably intense processes that were wholly distinct from those of today. Many geologists conclude that the Tapeats was originally deposited by a series of rapidly moving currents. This would have required a dynamic, regional catastrophe unlike any in modern experience.

A brief look at another fossil-bearing rock unit in Grand Canyon will illustrate these concepts. About halfway down the canyon walls is a prominent reddish cliff called the Redwall Limestone, found throughout the American West. Fossils are relatively rare and unimpressive in Grand Canyon, but in 1966, within a six-foot member in the center of the Redwall, about a dozen beautiful, two-foot-long, cigar-shaped nautiloids were discovered in a remote side canyon. Evolutionary uniformitarians have long believed this limestone layer was formed in a calm, placid sea, but creation geologist Dr. Steve Austin reasoned that the regional, dynamic water action of Noah's Flood was responsible. The fossil orientations are pointed in a preferred direction, indicating that moving water was involved. Exploring adjacent side canyons, he found other nautiloids where uniformitarians would never have thought to look. Identical fossils were eventually discovered in the same layer of rock throughout Arizona, Utah, Nevada, and New Mexico—an estimated four billion sea creatures. A catastrophic, widespread killer event had occurred. Surely, these remains represent an episode within the great Flood of Noah's day.

We have proposed that most rock layers and fossils,

Why Don't We Find More Human Fossils?

The fossil record abounds with the remains of past life. If the creationist interpretation of the fossil record is basically correct, most of the fossils were deposited during the Flood of Noah's day, as "the world that then was, being overflowed with water, perished" (2 Peter 3:6). These organisms were trapped and buried in ocean-deposited sediments, which later hardened into sedimentary rock.

But where are the pre-Flood human remains? According to Scripture, the patriarchs before the Flood lived long ages, and had large families and many years of childbearing potential. Where are their fossils?

First, we must consider the nature of the fossil record. Over ninety-five percent of all fossils are marine invertebrates with a hard outer shell, such as clams, corals, and trilobites. Of the remaining five percent, most are plants. Less than one percent of all fossils are fish, and even fewer are land animals. This encompasses reptiles (including dinosaurs), amphibians, mammals, birds, and humans.

In general, land creatures have a "low-fossilization potential." When land animals die in water, they float. It is very difficult to trap a bloated animal underwater in order for it to be buried. Furthermore, scavengers readily devour both flesh and bone. Seawater and bacterial action also have the potential to destroy organic remains. Most of the Flood's sediment deposition occurred at the ocean bottom, with debris carried along in violent underwater mudflows. The scouring ability of such mudflows, common during the Flood, would grind bone to powder.

The land-creature fossils that are found were mostly laid down during the Ice Age that followed the Flood. This was a land-oriented event that had the ability to bury animals in land-derived deposits. Human fossils have been discovered in such sediments.

The primary purpose of Noah's Flood was to destroy the land communities—especially humans—not preserve them. Some creationists even postulate that pre-Flood continents were subducted into the mantle, thus totally obliterating all remnants of human civilization. In any Flood model, the land fossils that were preserved would mostly have been buried late in the Flood. They would therefore be near the surface and subject to erosion and destruction as the Flood waters rushed off the rising continents.

Furthermore, there was overwhelming violence in the pre-Flood period (Genesis 6:13). Bloodshed and carnivorous activity no doubt terminated many family lines in both humans and animals.

For purposes of discussion, let us assume that 300,000,000 people died in the Flood and that their bodies were preserved as fossils that were evenly distributed in the sedimentary record, which consists of about 300,000,000 cubic miles of sedimentary rock. In other words, one human fossil was entombed in each cubic mile of sedimentary rock. The chances of such a fossil intersecting the earth's surface, resisting erosion, being found by someone, and then being properly and honestly identified is vanishingly small.

On the other hand, if evolution is true and humans have lived on the earth for three million years, many trillions have lived and died. Where are their fossils? This is the more vexing question.

with certain exceptions, resulted from the great Flood. If such a worldwide Flood were the cause, what would we expect to find? We should expect widespread deposits of a catastrophic character blanketing the continents. And that is exactly what we do find. More and more geologists are beginning to acknowledge that cataclysmic mechanisms are responsible for the earth's rock and fossil deposits.

The Makeup of the Fossil Record

Creation thinking predicts that fossils resulting from a dramatic marine destruction of the continents would be predominantly from sea-dwelling plants and animals that were trapped in oceanic currents that flooded the land. The upper sediments would increasingly reflect terrestrial habitats and inhabitants. This prediction has also been borne out, although pre-Flood environments were not necessarily the same as those seen today.

Pre-Flood ecology is envisioned as containing vast floating forests adjacent to coastal shorelines. These perhaps extended far out into the sea, with abundant shallow marine organisms living beneath the forest growth. As the "fountains of the great deep" (Genesis 7:11) broke open, initiating the Flood, denizens from the ocean depths were the first to be buried as sediments were pushed inland by dynamic currents. Eventually, shallow and upper-marine creatures succumbed, along with those in near-shoreline surroundings. As the Flood progressed, terrestrial dwellers and continental habitats were buried. The forces involved were mighty, totally annihilating many of the more fragile creatures. Those with hard outer shells possessed the highest fossilization potential, while those with poorly-connected extremities (like mammals) would rarely remain intact. Most mammals and birds would bloat, float, and be eaten by scavengers, while creatures that were submerged in sediment-dense waters would be abraded to powder by devastating underwater mudflows.

As the initial intensity of the Flood abated, land animals that had survived the first episodes finally succumbed and were buried in sediments containing a wide variety of fossils. Last of all, terrestrial fossils and environments were interred and plants floating on the surface of the waters were beached. Later stages of the Flood and post-Flood would have been marked by intense volcanism, particularly as mountains rose and continents spread. This generated great quantities of sediments for further burial.

The horrors of the Flood year ended for Noah and his family long before the earth quieted down geologically. All earth systems experienced readjustments during the following decades and centuries. These readjustments included the great Ice Age, during which weather patterns, jet streams, and ocean currents struggled to regain equilibrium. The fossils reflect that turmoil.

The discovery of a dinosaur or mammal fossil makes headlines, but these finds are extremely rare. When it is acknowledged that many of the existing mammal fossils were deposited in the turbulent times following the Flood, such as during the Ice Age, we realize why the mammalian fossil record of the Flood is so scanty. Creationists have predicted that we will not uncover many such Flood fossils, and, in fact, few are found.

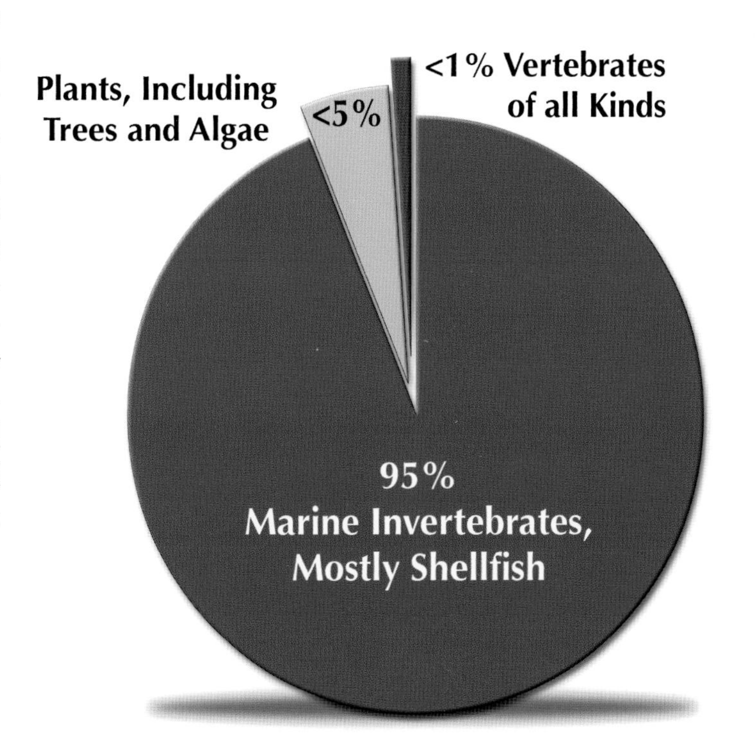

The great majority of fossils are the remains of marine invertebrates, nearly all found in catastrophic, widespread deposits on the continents. The geologic past of our planet was dominated by processes that were quite unlike those of today.

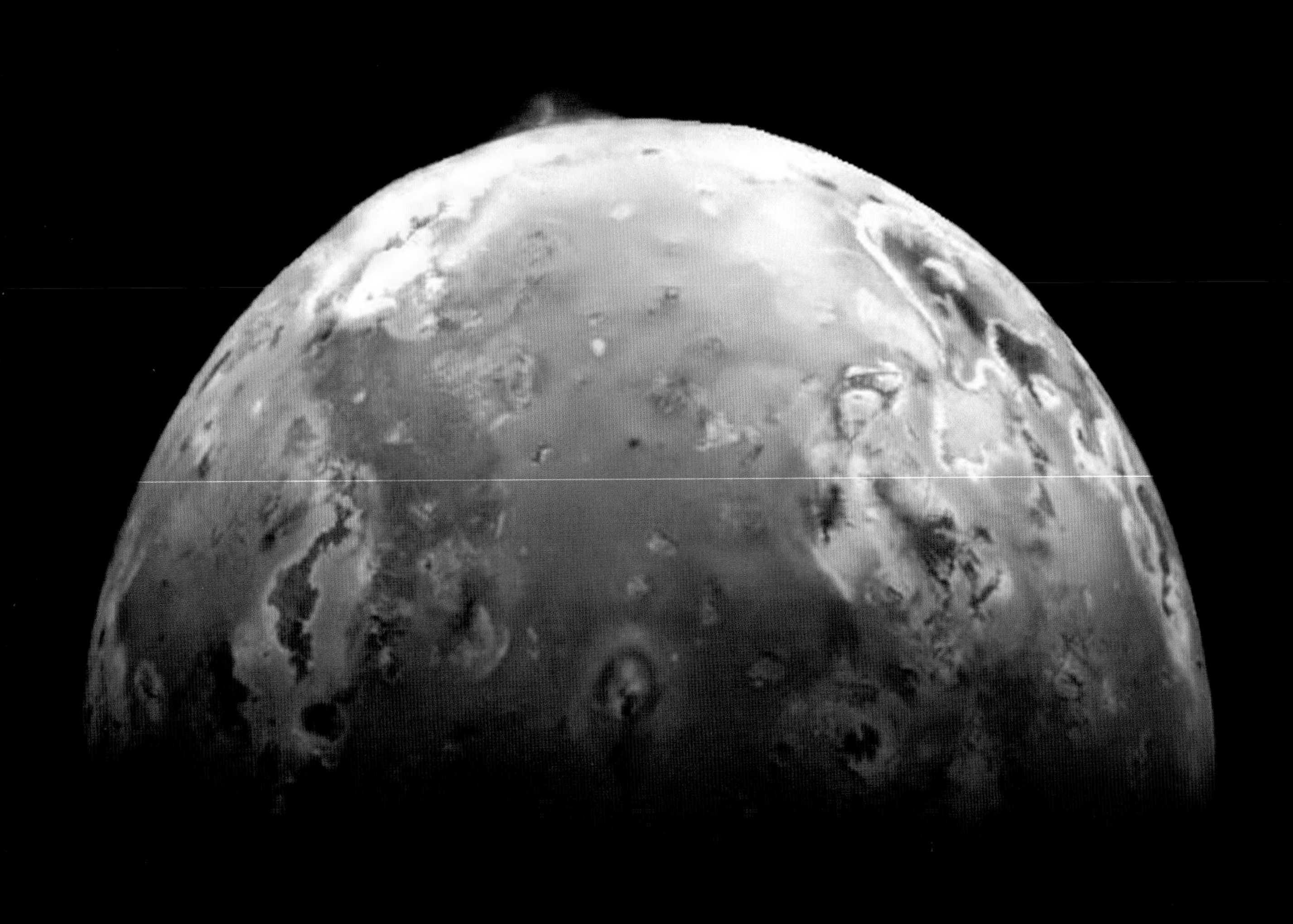

Io, one of Jupiter's moons. The desperate need to find out how life could have spontaneously generated has led evolutionists to look on other planets for suitable conditions. In our solar system, only the moons of other planets are now considered as possible sites. Of course, no such moon has been explored in person, and data from probes have not been promising. All locations are quite hostile for life. The search has degenerated into the search for water in any form and in any location. But finding water is not the same thing as finding life. Life is so complex, so engineered, and so intelligently designed that wherever it exists, it must have been created.

Chapter 6 Fossils of the Earliest Life

Whether the fossil record formed over months or millions of years, it is helpful to discuss it in the context of the passage of time and to demonstrate which fossil or fossil-containing layer came first in a relative sense. The standard geologic column or timescale is a conceptual device used by both creationists and evolutionists to help picture relative time. The standard column may have its flaws with respect to the absolute times assigned, but it can help us distinguish which geologic system with its fossils came before another, since events occurring earlier in the Flood (or earlier in evolutionary time, according to evolutionists) are recorded lower on the geologic column. Similarly, the term "abrupt or sudden appearance" for a plant or animal fossil involves a time concept, since it implies that a fossil type is found in one layer but not in any layer beneath (i.e., before) that.

The very earliest (or oldest) rocks, found at the bottom of the column, are crystalline, igneous rocks, which contain no fossils. From an evolutionary perspective, this is thought to be the time before the spontaneous generation of life.

Immediately overlying the fossil-free layers are fossils of the earliest forms of life. Fossils are said by evolutionists to increase in complexity and variety in the ascending levels of the column as time progresses. The following chapters will consider the dominant fossils in these layers, or "time periods," and whether the times they represent are astonishingly short or unimaginably long.

The idea that life arose spontaneously from non-living chemicals is easily the most difficult hurdle for evolution to overcome. Even the tiniest, simplest living cell abounds in complexity and biological engineering. Its intricacy seems far beyond the reach of random causes and statistical probabilities, and the concept that life self-generated through purely natural processes stretches credibility. Such a belief is actually incompatible with the level of organization seen in the DNA code. Once life exists, mutation and natural selection might act on it, and it might conceivably have a remote chance to evolve into something else—but there is no chance it originated on its own. The precisely arranged amino acids, proteins, and genes are direct evidence of creation by intelligent design. This is more than an overwhelming probabilistic argument against evolution. The signs of design are "clearly seen," and those who ignore them are "without excuse" (Romans 1:20). It is more reasonable—although remote in the extreme—to claim that the descendants of a living cell—however "simple"—evolved up through the chain of life all the way to modern man, than it is to insist on the first necessary evolutionary step from non-living chemicals to a living, reproducible cell.

This is why some modern evolutionary theorists have resorted to the proposition that life may have been seeded here on earth, either through a fortuitous happenstance when a life-bearing meteorite impacted the planet, or by unknown aliens who placed it here. This is also why so much hope is invested in the frantic search for conditions favorable for life (especially the presence of water) on other moons or planets.

Some treat the discovery of water on an extraterrestrial body as almost the equivalent of finding life. Water is necessary for living organisms, but chemicals would need much more than water to spring to life. In specified content, the organized complexity of life—complete with the information-loaded DNA code—overwhelms all the works ever written by man. If any scenario is impossible in science and illogical in thought experiments, the origin of life from non-life is it. But such a subject must be left to another treatment.

The Earliest Life

Life is thought by evolutionists to have generated in the Precambrian period (i.e., before—and thus below—the Cambrian strata) many millions and even billions of years ago. Fossils in the Cambrian characteristically have hard outer shells and other hard parts, and thus fossilized well. But in the evolutionary scenario, the multi-cell ancestors of the Cambrian fauna, as well as their own more simple ancestors, must have evolved from single-cell life with no hard parts at all. In their multi-billion-year odyssey, they must have gradually gained such parts, some acquiring exoskeletons or even shells, and thus been good candidates for fossilization. Certainly in over a billion years, the right conditions for preservation would have on occasion been met. The desired chain of ancestors, however, has not been found.

Previously, it was common for evolutionists to explain away the lack of early fossils by appealing to the unlikelihood of soft-bodied, fragile creatures surviving the fossilization process in recognizable form. This rationalization is no longer valid. We now know that soft-bodied animals have often been fossilized. For example, fossils of jellyfish (which are mostly made of water) have been discovered in great quantities. Furthermore, numerous techniques have been developed to coax tiny, ephemeral fossil remains—even microscopic bacteria—from their rocky tombs. We can tease out fossils of tiny cells with no nuclear membranes (prokaryotes) by etching away the host rock with acid. These can sometimes only be seen with a high-powered electron microscope.

Many find it difficult to believe that one-cell organisms (like those living today) could have descended into multi-cell ones, complete with primitive hard parts. Surely, we should have found some intermediate ancestors by now. We find stromatolites—thought by most to be mats secreted by single-cell blue-green algae or bacteria—in the lower strata, but they scarcely differ from modern algal deposits. These are certainly not "simple" organisms, for they harness sunlight to produce complex enzymes and sugars. Where can we see evolution in this process?

Unique or controversial "fossils" are often found, but evolutionists can no longer use an impoverished fossil record or inhospitable conditions as reasons for the lack of early evolutionary evidence. Ancestors for the varied multi-cellular life forms are, and will remain, missing.

The Cambrian Explosion

According to Darwin's "tree of life," non-life gave rise to one type (or no more than a very few types) of life, which then gave rise to all the others. Thus, the tree sports only one "trunk," with multitudes of branches and twigs that developed over time in a tree-like pattern from bottom to top. In contrast to this depiction, the Cambrian portion of the fossil record has preserved multitudes of invertebrate types that all appeared at the same time, each quite complex and quite different from the others. An honest look at when and where fossil types are found suggests that life began with a multitude of early life plans, not with a single plan that later branched out. The fossil record looks more like a lawn than a tree.

Note that this is quite unlike the Darwinian predictions, wherein species slowly and continually gain new traits and features, and proliferate into side branches that eventually become separate types themselves.

> This rapid diversification, known as the Cambrian explosion, puzzled Charles Darwin and remains one of the biggest questions in animal evolution to this day. Very few fossils exist of organisms that could be the Precambrian ancestors of bilateral animals, and even those are highly controversial.[1]

The layers immediately overlying the Precambrian-Cambrian boundary contain a stupefying abundance

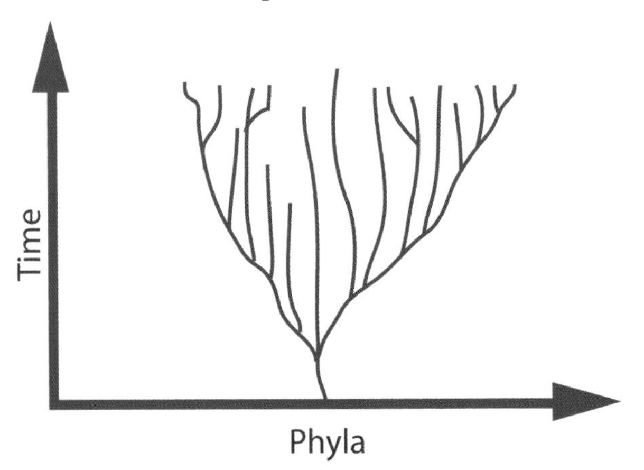

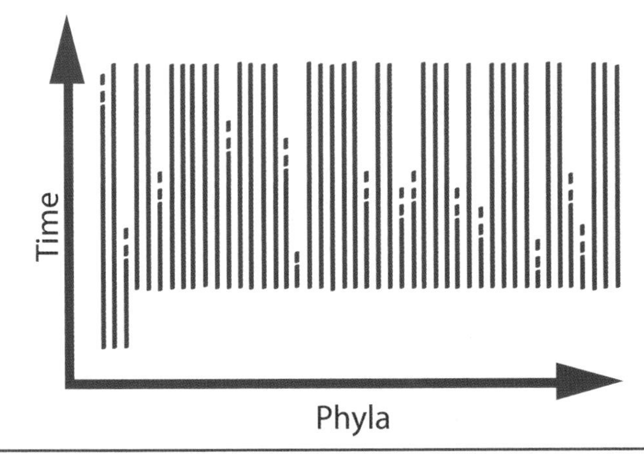

Evolution proposes that only one type (or a very few types) of life formed by spontaneous generation from non-life. It holds that life's diversity increased over time into many basic body styles. Creation thinking insists that essentially all living creatures appeared suddenly and basically retained the same forms (stasis) through time. Some types have gone extinct over time, but no new basic types have arisen.

Fossils from the Ediacara Hills of Australia are now dated as Late Precambrian. Actually, one could consider the specific date assigned to be Early Cambrian, with the boundary moved, but in the end it does not really matter. The Ediacaran fossils are not ancestral to those seen in the Cambrian explosion of life. They are quite different, and no affinities can be discerned. Now there are more phyla than ever that are unaccounted for in the evolutionary interpretation of early earth history.

of life, including virtually every basic animal type from marine invertebrates to vertebrate fish. This abundance is aptly dubbed the "Cambrian explosion" of life. Many of these Cambrian animals are quite like extant (living) animals, and are exquisite in their complexity. They suddenly appear, simply springing to life (as recorded in their fossils, of course) fully formed, without a hint of ancestral types.

Cambrian trilobites abound, with eyes at least as proficient as those possessed by any animal living today. Many creatures—like clams, brachiopods, starfish, and cephalopods—possess hard outer shells, but others are soft-bodied, like jellyfish. An evolutionist would reasonably expect the preceding Late Precambrian layers to contain fossil remnants of their ancestors.

> As Darwin noted in the *Origin of Species,* the abrupt emergence of arthropods in the fossil record during the Cambrian presents a problem for evolutionary biology. There are no obvious simpler or intermediate forms—either living or in the fossil record—that show convincingly how modern arthropods evolved from worm-like ancestors. Consequently there has been a wealth of speculation and contention about relationships between the arthropod lineages.[2]

Discriminating readers will note something different about the chart on page 28 from those normally presented—an Ediacaran period has been added to the uppermost level of the Precambrian Era. In the Ediacara Hills of Australia and a few other places, strange fossils have been discovered below the base of the Cambrian layers. The strata that contain them cover an area of lesser extent and in relatively fewer locations than other fossil-bearing strata. These are touted as possible Precambrian ancestors for the Cambrian fauna, but since no fossil can be directly dated, their place in the geologic column is uncertain. Were these from the lower Cambrian, or were they deposited before the Cambrian? After much contentious deliberation, it was decided they fit just below the lowest Cambrian in the underlying uppermost Precambrian time, and that they were indeed in position to be ancestors of Cambrian animals.

Paleontologists had previously reported controversial hints of indirect trace fossils, like worm burrows, and even possible multi-cell organisms in Precambrian layers, but their character and date were questionable. Impressions that suggest jellyfish, segmented worms, and odd, soft-bodied invertebrates similar to the Ediacaran fossils have now been discovered in several places around the world. They are considered to be older than the Cambrian, but the specific dates are still under debate. Were they from Early Cambrian or Precambrian?

Most Ediacaran fossils are of unknown affinity, with no modern counterparts. It is difficult, if not impossible, to arrange their family tree. There are some that vaguely resemble known types of jellyfish, worms, and coral alive today, but the rest are—how shall we say—bizarre. Nearly all are soft-bodied, yet fossilized. They simply are not candidates for the array of recognized body plans that burst onto the scene unannounced in the Cambrian explosion. Nevertheless, in 2004 these disparate creatures were given their own period on the revised geologic column (the Ediacaran period), and the Cambrian was shortened.

But this is still no help to the evolutionist searching for support for his theory. Whether the boundary is called Precambrian–Cambrian or Ediacaran–Cambrian, the Cambrian explosion of complex life is a verified, dominant feature of the fossil record. This presents more of a problem for Darwin's modern followers than it did for Darwin himself. Whether they are called Precambrian or Ediacaran, the fossils found lower than the Cambrian seem to suggest that the ancestors of the Cambrian organisms simply did not exist. Otherwise, they would have been fossilized. Sudden appearance is the rule.

The best evolutionary interpretation of these unusual Ediacaran fossils would be that they are the as-yet unrecognizable incipient stages of Cambrian fauna. This, however, would be a statement of faith, not science. The anticipated claim that these eerie things were the fossilized ancestors of complex invertebrates cannot be made. They were, in fact, complex invertebrates themselves, whether or not they are familiar to us. Their lines seem to have gone extinct.

Darwin was convinced that the fossilized ancestors of the Cambrian fauna would someday be found. He predicted they would eventually be discovered and solve the mystery. It has been over a century and a half since Darwin's prediction, with multitudes of enthusiasts scouring the planet for transitional fossils. Museums are brimming with fossils, yet the only ones claimed to be transitional are highly controversial, even among evolutionists. Museum cases should be full of transitional fossils, with body parts that are not yet fully evolved. Where are the in-between forms? Can we still hold out hope that they will turn up? As we shall see in a later discussion, the fossil record can be considered essentially complete. Darwin's worst fears have been realized.

Possible Creationist Explanation

Evolutionists rightly attempt to fill in the gaps in their theory, but there is very little material for them to work with. Since the early fossils do not fit the standard evolutionary story very well, it behooves us to look at the creationist explanation.

In general, where there is abundant fossil data, there is no hint of evolution. When clam fossils, for instance, are found, there are no identifiable ancestors present. Clams are clams. They are found by the trillions, in many different layers. They may vary, but they remain recognizably clams. When just a few fossils of a creature are found, only then can a controversial case for evolution sometimes be made.

According to Genesis, the world was created as an incompletely formed, uninhabited place. Perhaps it consisted of nothing but a rotating slurry of watery matrix called "the deep" on Day One (Genesis 1:2). But then the earth underwent a series of God-ordained modifications in preparation for life. On Day Two, the oceans and atmosphere were separated. On Day Three, the continents were raised, preparing the planet for life. The same day, plants were created as eventual food for the animals and man. On Days Five and Six, animal life appeared in the sea and sky, and then on land. Mankind was then given stewardship responsibility over creation, under God's authority.

Note that there was a time before plant and animal life was created when geologic processes operated and may have deposited non-fossiliferous sediment. Waters rushing off the uplifting land surface on Day Three would have eroded parts of the new continents, depositing sediments on the sea floor. These may have trapped and buried certain non-"living" organisms.

Between Adam's sin and the great Flood, rivers flowed, erosion proceeded, and sediments accumulated. We could expect certain small, non-mobile organisms to have been trapped and deposited in the sediments that resulted from the rather calm processes during this time period—processes not wholly unlike what uniformitarian evolutionists have envisaged operating during the four billion years they assign to earth's history. This relative tranquility was then shattered completely in the great convulsion of earth in the days of Noah.

Thus, the early, pre-life, completely fossil-free deposits would be overlain by easily-transported, tiny traces of early "life" in pre-Flood offshore deposits. These would eventually be overlain by deposits from the great Flood, teeming with animal and plant remains from all habitats. Thus, in general, the creation/Flood model predicts the fossil sequence at least as well as does the evolution model.

The Fossil Record's Completeness

The fossil record's imperfections have been offered as

Six Days of Creation
DAY 1
DAY 2
DAY 3
DAY 4
DAY 5
DAY 6

the reason that so few intermediate fossils have been found. It is claimed that only a minor percentage of transitional forms would have been fossilized, so why should we expect to find all types? Darwin himself entertained such notions: "The explanation lies, as I believe, in the extreme imperfection of the geologic record."[3] But does this reflect reality? Does it solve the problem?

For the answer, let us turn to the record itself and see just how well the various categories are represented. First, we must recall just how many types of living things have existed and do still exist. Life's vast variety is catalogued in groupings based on similarity—first in basic body plan, and then in lesser characteristics. The overarching classification in the taxonomic system is by kingdom, of which there are five (or three, depending on the preference of the individual researcher), including plants and animals. Interestingly, all kingdoms are found as fossils from near the bottom of the column to the top, and still have living representatives today.

Considering the animals only, the next biggest grouping is the phylum, referring to the basic body plan of an organism. As already mentioned, every one of the 30 to 50 phyla (depending on the researcher's criteria) has been discovered among the fossils of the Cambrian explosion. Some individual phyla have gone extinct, but most are found throughout the record and still exist today. Where are the record's imperfections?

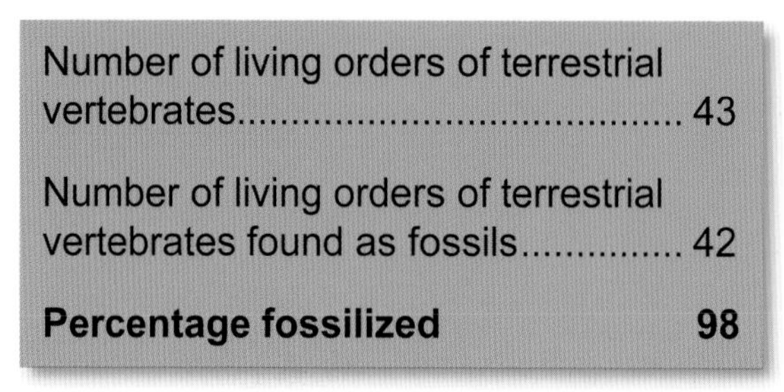

Number of living orders of terrestrial vertebrates	43
Number of living orders of terrestrial vertebrates found as fossils	42
Percentage fossilized	**98**

Number of living families of terrestrial vertebrates	329
Number of living families of terrestrial vertebrates found as fossils	261
Percentage fossilized	**79**

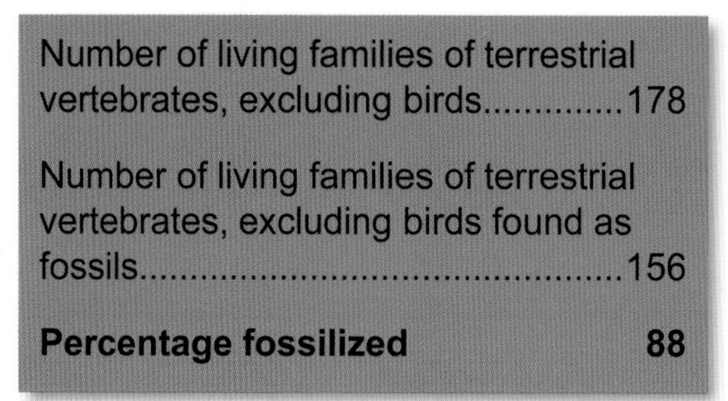

Number of living families of terrestrial vertebrates, excluding birds	178
Number of living families of terrestrial vertebrates, excluding birds found as fossils	156
Percentage fossilized	**88**

Adapted from Denton, 1985, 190, Figure 8.5.

Much the same could be said for the classes, the next lower category. Of the (usually poorly preserved) terrestrial vertebrate classes, representatives of each are found as fossils.

Following this are the classifications of order and family. In his illuminating book *Evolution: A Theory in Crisis*, Dr. Michael Denton used the listings in Romer's 1966 classic work *Vertebrate Paleontology* to compile the percentages of living orders and families that had been found in fossilized form.[4] Certainly many more fossil families and orders have been uncovered since 1966, so the percentages should be higher. The only conclusion one can draw is that the fossil record is virtually complete with respect to basic types. Such a high percentage of taxonomic fossil groupings reveal that the only ones lacking are the transitional fossils. The links that were missing in Darwin's day are still missing.

The presence of gaps among fossils poses a difficult enough problem, but it is magnified when we recognize that the gaps are systematic. They mirror the differences between the "kinds" found in the creation model and are exactly the same gaps that appear among living types.

The regular, systematic nature of the fossil gaps also speaks in another way. Evolution is normally touted as proceeding in small steps, with new genes and new traits accumulating by mutation through generations and over millennia. Conceptually, many more steps are required to fill a large gap than would be needed for a smaller gap. Thus, there should be more intermediate representatives and potential fossil candidates. Today's phyla, identified according to their fundamentally different basic body plans, should be connected by abundant transitions. But very few are claimed to have been found, and in reality, essentially no candidates exist. Did evolution occur?

Modern species can be arranged to make an evolutionary statement. A reptile can be placed midway along a straight line between an amphibian and a mammal, but where is the documentation that the amphibian participated in such a transition? How do we know a reptile descendant became a mammal at all? All we have are separate, distinct organisms, placed in an artificial arrangement. An evolutionary tree so constructed is not a documentation of how evolution occurred, but an array of end point data in a possible evolutionary scenario.

In creation, each animal type suddenly appeared at the same time as the others, and did not descend by modification from another animal. Much variation was allowed. Those species which are varieties of a created "kind" might be linked by transitional fossils, but none exist between the basic categories. These are the gaps we see in the fossil record, just where they ought to be if creation is true.

Let us not allow the lack of transitional fossils to obscure the real problem paleontology presents to evolution. The record from the past may be silent on whether or not evolutionary change occurred. However, we need not simply focus on the lack of data to support evolution, but on the thundering message of the fossil evidence, which comes through loud and clear. The fossils speak clearly of the sudden appearance of abundant complex kinds, separate from others, just as Scripture claims. This was followed by stasis, also just as implied by Scripture.

A famous "living fossil." Trees showing a fan shape in the veins of its leaves were thought to have gone extinct over one hundred million years ago, until a living community of ginkgo trees was discovered alive today.

This beautiful crinoid fossil was buried quickly, before the fragile animal could be broken apart. It looked much like a modern sea lily, but is classed as an animal. It lived anchored to a stable surface and fed on passing nutrients in the water. (Location: Indiana. Dated as Mississippian.)

Chapter 7 Origin of the Invertebrates

The first big problem for evolution is the origin of life from non-living chemicals, with neither organic mutations to act as sources for genetic novelties nor natural selection to act as a filtering guide. There was no life anywhere before this, thus no food, no potential mates, no competition, only unforgivably harsh conditions. Natural selection could not help, for there was no variety of life from which to select. But assuming that single-cell life did have an inorganic start, the next big problem for evolution is the origin of invertebrates (which are, in and of themselves, amazingly complex) from single-cell precursors.

The Cambrian explosion exemplifies the problem, with the sudden appearance of virtually every phylum and no ancestral lineage for any of them. The discovery of the Ediacaran assemblage does not really solve the problem, for these creatures were fully formed, very complex, and also do not have ancestors. If anything, they make the situation worse and evolution's mountain becomes even harder to climb.

Through what stages did one-cell life develop into invertebrates? Which single-cell ancestor gave rise to the clams? Which gave rise to the corals? The sponges? The arthropods? Each invertebrate type stands apart from the others, each is complex, and each is fully fit for its environment. What is the origin of any of them?

Evolutionist Robert Barnes was candid regarding the origin of invertebrate animals when he said:

> The fossil record tells us almost nothing about the evolutionary origin of phyla and classes. Intermediate forms are non-existent, undiscovered, or not recognized.[1]

With the abrupt appearance of so many different invertebrates, none with known ancestors, a related conundrum arises. Since all types arrived essentially simultaneously (as recorded in the Cambrian fossils), with each phylum distinct from the others, none could be the ancestor of any other. Nor is there any fossil evidence of relatedness. What was happening in nature to force so many innovations at the same time, and from what did they descend? Truly, Darwin's tree of life does not adequately model reality.

> If ever we were to expect to find ancestors to or intermediates between higher taxa, it would be in the rocks of late Precambrian to Ordovician times, when the bulk of the world's higher animal taxa evolved. Yet transitional alliances are unknown or unconfirmed for any of the phyla or classes appearing then.[2]

The Burgess Shale

One extraordinary fossil find is often mentioned in this context. In 1909, fossils of amazing clarity and diversity were found in Canada's Burgess Shale by Charles Walcott of the Smithsonian Institute. The extremely fine-grained shale preserved intricate details of previously unknown invertebrates, including everything

Exquisite fossils found in Canada's Burgess Shale Formation have been touted by some as the transitional fossils that evolutionists so desperately need to account for the Cambrian explosion of life. But this is not so. These fossils are unique and are not ancestors of the Cambrian fauna. They actually have made evolution's problem worse, since they add to the list of phyla without ancestors. Furthermore, the date assigned to this formation is mid-Cambrian, not Precambrian or Early Cambrian.

from their hair and bristles to tiny external structures. It even preserved hints of the nature of their internal organs.

Does the Burgess Shale contribute to our understanding of invertebrate evolution, as is often claimed? It did yield much useful information, unveiling numerous unfamiliar fossil types, as well as exotic species of phyla that were already known. Depending on the authority consulted, perhaps several new phyla were also found. Specimens might have been different enough to warrant a new phylum designation, but none of the fossils were primitive in any sense, or seemingly ancestral to any other type. What the Burgess Shale fossils certainly did not do was reveal any of the transitional forms evolution so desperately needed.

This formation has been dated as Middle Cambrian, only slightly "after" and stratigraphically above the strata representing the initial Cambrian explosion. As a result, there are now even more basic types in the lowest system than had been known before. Adding to the number of phyla that must be accounted for does not help solve evolution's huge problem.

Consider the magnitude of this puzzling situation from an evolutionary perspective. Complex, multi-cell life sprang into existence without adequate ancestral forms. Not one basic type or phylum of marine invertebrate is supported by an ancestral line between single-cell life and the participants in the Cambrian explosion. Nor do the basic phyla seem to be related to one another. How did evolution ever get started?

The creation model predicts that no ancestral forms

This clam was buried alive in a pose of self-preservation, with its two halves tightly shut for protection, not in a normal living position. It was quick-buried in a fossil graveyard that was composed of myriads of marine invertebrates.

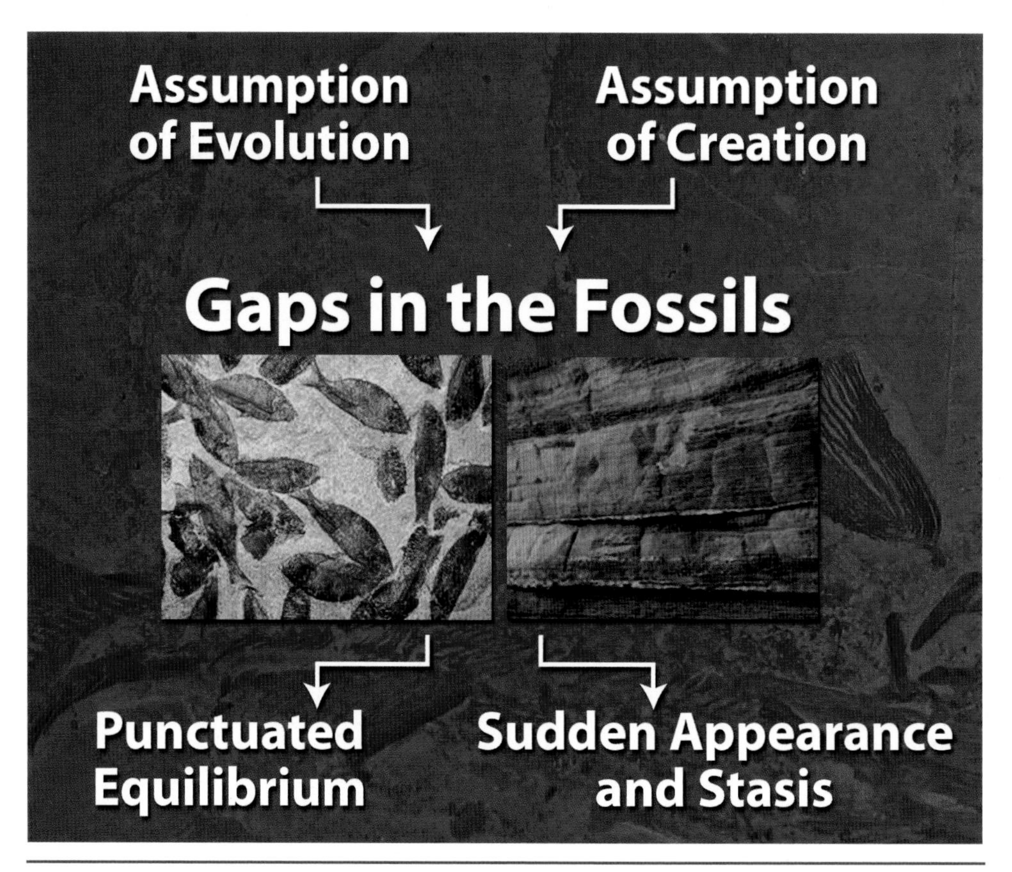

How can evolutionary and creation scientists arrive at such divergent interpretations of the same data? The beginning assumptions held by each group are the reason. To a great degree, these assumptions dictate the conclusions that are reached.

will ever be found, for they never existed. Each basic body plan was created, without any ancestors, directly from the mind of the Creator. It does no good to claim ancestral forms were not preserved or are not needed, for their absence directly matches the predictions of creation. Certainly if the transitional forms had been found, they would be paraded for all to see. Creation/evolution discussions would be welcomed in the science classroom, rather than the current censorship of any criticism directed against evolution. Supported by overwhelming evidence of relatedness between life's present forms and less complex ancestral forms, there would be much less doubt about evolution. It would indeed be hard to maintain creation views. But there is no such evidence.

The creation model also predicts the sudden appearance of life. The data show exactly what they should show if creation is true. Evolution must explain its lack of supporting data, or modify its model to fit.

"Punctuated equilibrium" is often touted as a solution to the problem of missing transitional forms. Popular on university campuses since the 1970s, this modification of evolutionary theory strives to "explain" the lack of supportive fossil data. It proposes that life forms existed in equilibrium in an unchanging environment for long periods of time without the need for change, and thus remained the same. When faced with a sudden change in their environmental conditions, they responded by rapidly acquiring new traits—so rapid, in fact, that they left no fossils, and thus no record of the transitional stages of the change.

Many insist that evolution operates too slowly in the present to be observed. True enough, we do not see it occur today. But according to punctuated equilibrium, it occurred so fast at times in the past (in evolutionary terms) that it left no trace. How convenient that the transitional forms that would substantially "prove" evolution left no fossil trace. This argument from the lack of data may be consistent with the message from the fossil record, but it is not the only conclusion that can be drawn from the evidence. And it is certainly not the best conclusion. The fossil record is more consistent with the creation model's insistence that such fossils never existed and will never be found.

The record of the fossils makes it impossible to determine the ancestry of any basic group of plant or animal. As mentioned earlier, clam fossils are found all over the world, in rocks of every age, but no ancestral

A round "river rock" sometimes contains a fossil inside. When tapped, it splits open and the fossil appears. This marine fossil was found on Mt. Everest, the tallest mountain in the world. What is a marine fossil doing on Mt. Everest, you may ask? The great Flood of Noah's day formed many sedimentary layers on the ocean floor that hardened and later buckled up into mountains.

A mass burial of trilobites. These free-swimming and crawling marine invertebrates normally would have no difficulty surviving oceanic turmoil. But these were buried en masse, crowded together, by a major geologic event. Trilobites are found throughout the early Flood record. Their different forms were simply varieties of the trilobite kind. Since they appear to be extinct, we may never know for certain. (Location: Morocco. Dated as Cambrian.)

forms have ever been discovered. An evolutionary lineage is impossible to discern, for clams have been clams ever since the Cambrian explosion of life. Great variety among the clams can be seen, but variety does not speak of ancestry.

Clams and other invertebrates are found in many settings, including the tops of high mountains such as Mt. Everest, the tallest mountain in the world. Here, ocean bottom strata have been buckled up to a high elevation. Within the rock are clams, different from some modern varieties, but still recognizably clams. The same is true of all the different animals found in the Cambrian explosion. How can evolutionists use the fossils as evidence of a common descent of all life?

Ammonite fossils are among the most common fossil. They were cephalopods, and are represented today by the beautiful Chambered Nautilus, which is essentially a squid with a coiled shell. Their shells were adorned with "chambers" that formed during growth periods. Each growth period was usually about a month long, allowing the age at death to be estimated. In size, ammonites varied quite a bit. They could range up to six feet or more in diameter, but were commonly only a few inches or less in diameter. Their fossils are found in many locations in strata that are conventionally dated from the Paleozoic to the Tertiary.

Ammonites also varied widely in shell design, and evolutionists attach great significance to these differences. The fact that some ammonites possessed "ribbed" shells and others had smoother shells is taken as "proof" that macroevolution occurred in the past. Creationists are much less impressed by these minor changes, however, ascribing them to mere variation within the ammonite kind. The discovery of their fossils throughout much of the geologic column testifies to a remarkable stasis.

Not every suitable rock stratum contains fossils. Often rocks that might be expected to house abundant fossils have none, particularly those fossils that can be seen without magnification. "Fossils are where you find them," fossil hunters often say.

At any rate, the transitional fossils predicted by evolution have never been found. From all we can tell, transitional missing links are imaginary, necessary only to support belief in the evolution model.

To summarize, the origin of marine invertebrates from single-cell life is a huge problem for evolution. Appearing abruptly as they do, in great variety, with no transitional forms between them, and with no ancestors, they make a strong case for creation and aggressively argue against evolution.

A cluster of ammonites buried together. These animals were able to swim freely in the open ocean. We don't know whether they grouped together in life, but they died that way, in a watery catastrophe, crammed in with other creatures. (Location: Russia. Dated as Cretaceous.)

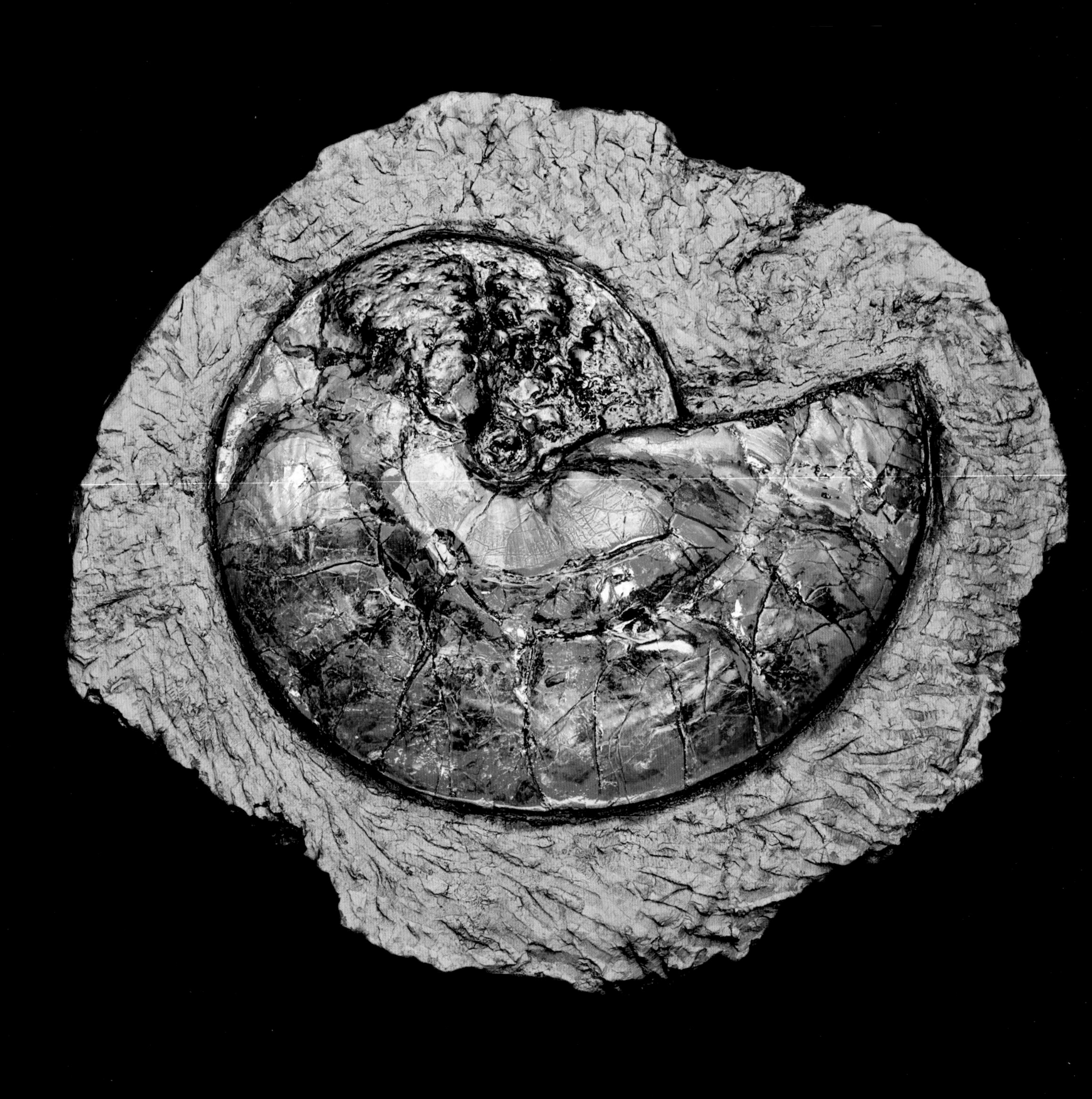

Ammonites are common fossils. Occasionally, conditions subsequent to their burial infused them with minerals and turned them into veritable gemstones. (Location: Bearpaw Formation, Canada. Dated at 70 million years.)

Chapter 8 From Invertebrates to Vertebrates

Another huge problem for evolution to overcome is the origin of the vertebrates from the invertebrates. What animal gave rise to the vertebrate fish—and, by extrapolation, the amphibians, the reptiles, the birds, the mammals, and finally man? Was it clams, or starfish, or sponges? Each invertebrate phylum is distinct from all others, and each is quite different from fish and the other vertebrates.

Three Big Problems for Evolution

Remember that many of the invertebrates possess a hard outer shell and are abundantly represented in the fossil record. Fish fossils are well-represented, fossilized by the billions, while the hard-shelled invertebrates are fossilized in even greater numbers, likely in the trillions.

Evolution assumes that vertebrates evolved from non-vertebrate chordates, a known group of animals with a stiff notochord down the back. But what is the evidence that they fathered the vertebrates, and from which invertebrate did the chordates come? Such an animal might be expected by evolutionists, but there is a huge difference between it and any vertebrate. Could it really be transitional between invertebrates and vertebrates? And how would we know?

What gave rise to the vertebrates? If it was a marine chordate, how did it evolve a backbone and become a fish, and then in turn develop into all the other vertebrate types, including man? If evolution cannot overcome this hurdle, it is (please pardon the expression) dead in the water.

The First Fish

For many years, complete fossils of fish have been known from Ordovician and Silurian layers. In recent years, they have been discovered in Cambrian strata, and have even been reported in the lowermost Cambrian layers—i.e., the Cambrian explosion itself. This "exploded" a key prior belief held by evolutionists. Cambrian fish fossils are typically incomplete, and many are from unusual fish, to say the least, but they are fish nonetheless.[1]

These fossils belong to the category of jawless fish, and often have bony scales for skin. Decades ago, only a few unattached bony scales were known from the Cambrian. Complete fossils have now been found on several continents, most notably China, revealing their bony owners to have been fully fish. Representatives of these strange fish types are alive today—armored fish that are thought to be reminiscent of the ancestors of all later fish. As to their own ancestry, nothing is known.[2]

Speculation among evolutionists for the origin of vertebrates from invertebrates now centers on the unusual creature *Pikaia*, nearly identical to the living creature *Amphioxus*. Did an echinoderm like a starfish really transform itself into a fish through this intermediate step?

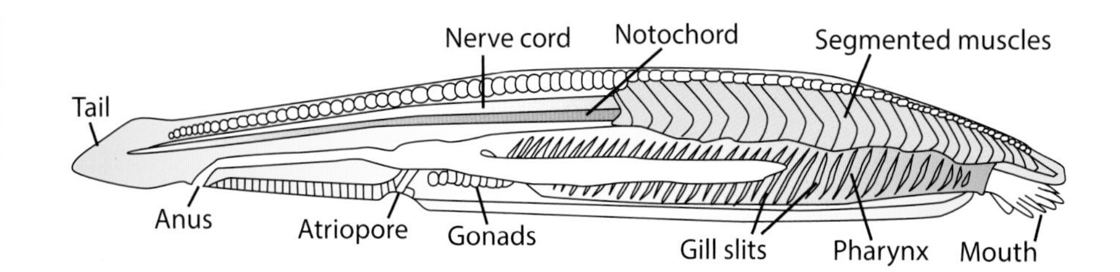

Diagram of *Amphioxus*, which resembles the extinct *Pikaia*, thought by some evolutionists to be an ancestor of modern vertebrates.

Many have speculated that Early Cambrian vertebrate fish evolved from a chordate creature that came from some invertebrate.[3] In this scenario, the evolutionary radiation of the chordates from their primitive relatives must have occurred rapidly within the Early Cambrian, or possibly before. Speculation now centers on a chordate similar to the modern creature *Amphioxus*, which has many features that are quite unlike those of fish. It has a central nerve chord running the entire length of its body, supported by a stiff rod-like structure, but it lacks a true backbone and a recognizable brain, or even a head. It also does not have teeth or a heart. This unusual non-vertebrate chordate is quite different from all vertebrates (as well as any invertebrate), yet evolutionists say that vertebrates evolved from some animal like this.

A fossil named *Pikaia*, nearly identical to the living *Amphioxus,* has been discovered in the Middle Cambrian Burgess Shale. It has no obvious ancestors among the other phyla, yet its theoretical descendants are legion—indeed, the entire vertebrate group. Evolutionists had loosely predicted that some animal similar to *Pikaia* must have existed during the period before fish appeared, but this fossil specimen lived after the appearance of fish in the Early Cambrian. Remember, the modern *Amphioxus* is nearly identical to *Pikaia*—a true "living fossil." As a matter of faith, one can believe that its kin lived prior to the Middle Cambrian. But that means that its undiscovered earlier form would have needed to adapt quickly from its stable but primitive form into different types of Early Cambrian fish. Note that the best candidate for an invertebrate changing into an early vertebrate ancestor has not changed at all in the presumed half-billion years since. A skeptical observer might marvel at how evolutionists use these organisms to claim a victory in their search for the ancestral vertebrate.

Which non-chordate gave rise to the chordates? Many modern theorists have proposed that chordates descended from echinoderms, up through the acorn worms (tiny, burrowing, worm-like creatures) and sea squirts (soft-bodied, sac-like filter feeders). Frequent speculations center on the echinoderm starfish, based on the similarity of the gut lining in the embryonic starfish to the gut lining in some vertebrates during embryonic development. This is indeed similar, but is it an indication of ancestry? In the adult forms, there is no such similarity.

> *Amphioxus* and a menagerie of other odd-looking creatures form a link between the invertebrate world, especially the echinoderms, and the vertebrates....The mesoderm (the embryonic gut lining) in starfish, *Amphioxus*, and humans similarly forms as an outgrowth from the gut wall.[4]

Evidence from embryonic similarities may be interesting, but without a heartfelt commitment to evolutionary relatedness, no ancestral relationship would be inferred. Remember, we are attempting to find evidence of evolution itself. Is it proper to use the assumption of evolution to prove evolution?

But evolution must choose a champion. From what invertebrate did the vertebrates arise? Is the best candi-

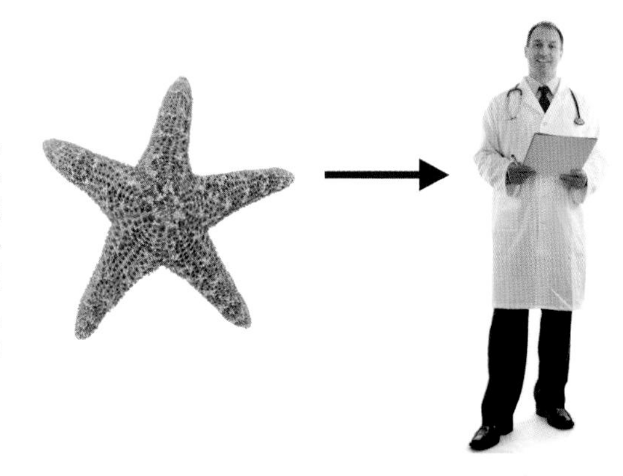

During embryonic development, starfish and other echinoderms employ a certain chemical compound in the gut area that is somewhat similar to a chemical utilized in the gut of vertebrate embryos. The proposal has thus been made that echinoderms such as starfish evolved into vertebrates. This includes all fish, amphibians, reptiles, birds, mammals, and man. Is this the best evidence evolution has for such an important transition? Unfortunately, yes.

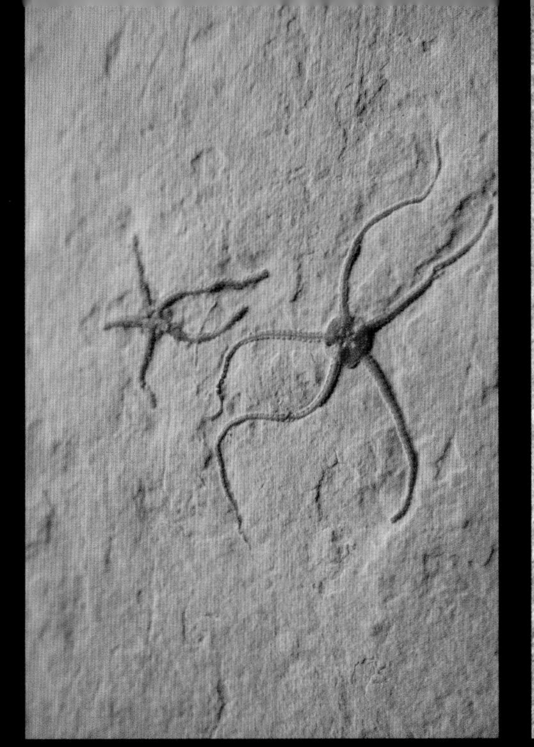

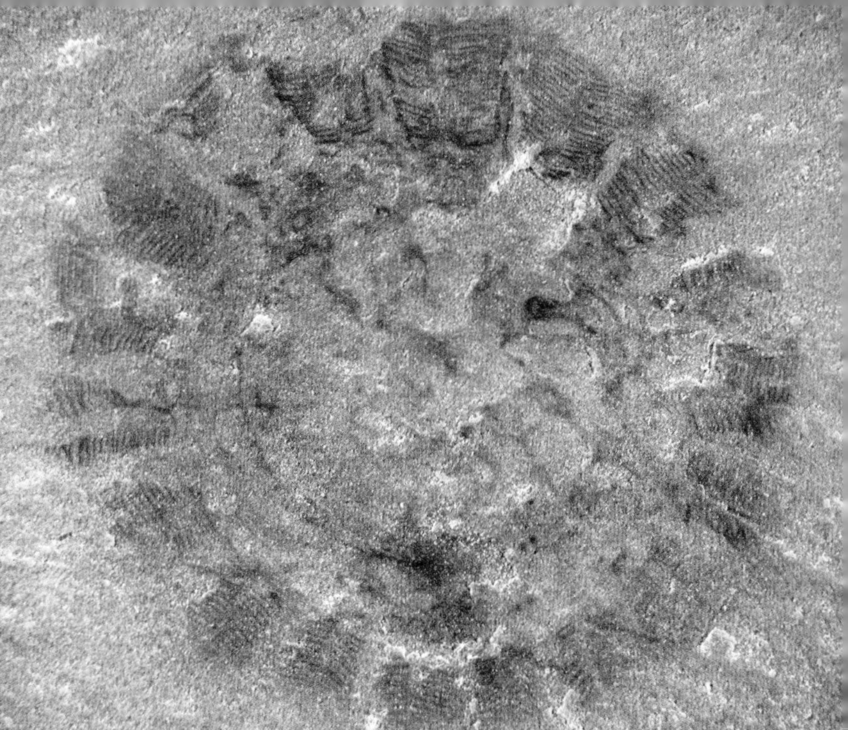

Which invertebrate evolved into the vertebrates? All phyla (including the vertebrate fish) are found in the Cambrian explosion of life, located near the bottom of the geologic column, and yet the origin of animals with a backbone remains an unsolved mystery. The dazzling array of invertebrates—from jellyfish to squid to starfish to snails—should document this important transition. But it does not.

The squid fossil at the bottom displays exquisite design. Scarcely different from modern squid, it demonstrates stasis, not evolutionary change. Likewise for the jellyfish. This specimen is the "oldest" one ever found, yet it is recognizably a jellyfish. In the upper left a starfish (or, more properly, a sea star), shows the typical five-arm design of echinoderms. Question: What would be involved in changing a sea star into a man? This could not happen through random mutation and natural selection in a billion years, or in a trillion years! Waving a magic wand of time over an impossible task does not make it possible. (Squid location: Germany. Dated as Jurassic. Jellyfish location: Utah. Dated as Cambrian.)

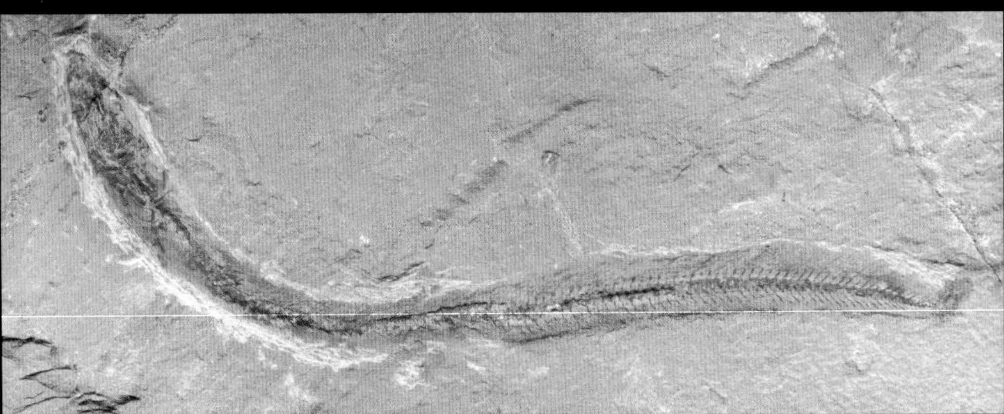

The Creator's account of creation includes the information "God created...every living creature that moveth, which the waters brought forth abundantly, after their kind" (Genesis 1:21). Surely, the oceans teem with life of an amazing variety. From recognizable fish, as in the lower right, to the filter-feeding paddlefish at the top (both from Green River, Wyoming, and dated as Eocene), to the Mesozoic eel on the lower left. Each fossil shows complex design and stasis over time.

date the starfish? An amazingly rich fossil record highlights the differences, not the relatedness.

Are Fish Categories Related?

> The key episodes of early fish evolution seem to have taken place during the Ordovician and Silurian (443-417 Mya), when all the major groups appeared.[5]

For the evolutionist, the rise of the vertebrates began with primitive fish, which we now know "suddenly appeared" in the Cambrian explosion. The next step is the evolution of the remaining fish categories. Has the paleontologist been any more successful at this task? Absolutely not. In fact, even less so.

> The higher fishes, when they appear in the Devonian period, have already acquired the characteristics that identify them as belonging to one or another of the major assemblages of bony or cartilaginous fishes....The origin of all these fishes is obscure.[6]

Particularly embarrassing for evolution is that fossils of bony fishes are found long before (i.e., in much lower strata) than those of cartilaginous fishes (sharks and rays), even though most evolutionists continue to believe cartilage came first and evolved into bone.

> All three subdivisions of the bony fishes appear in the fossil record at approximately the same time....How did they originate? What allowed them to diverge so widely?...And why is there no trace of earlier, intermediate forms?[7]

> But there is still much that is not known about the relationships among major clades of fishes.[8]

Given the lack of consensus on fish origins, it should come as no surprise that competing stories are frequently told. As a matter of fact, for any three objects

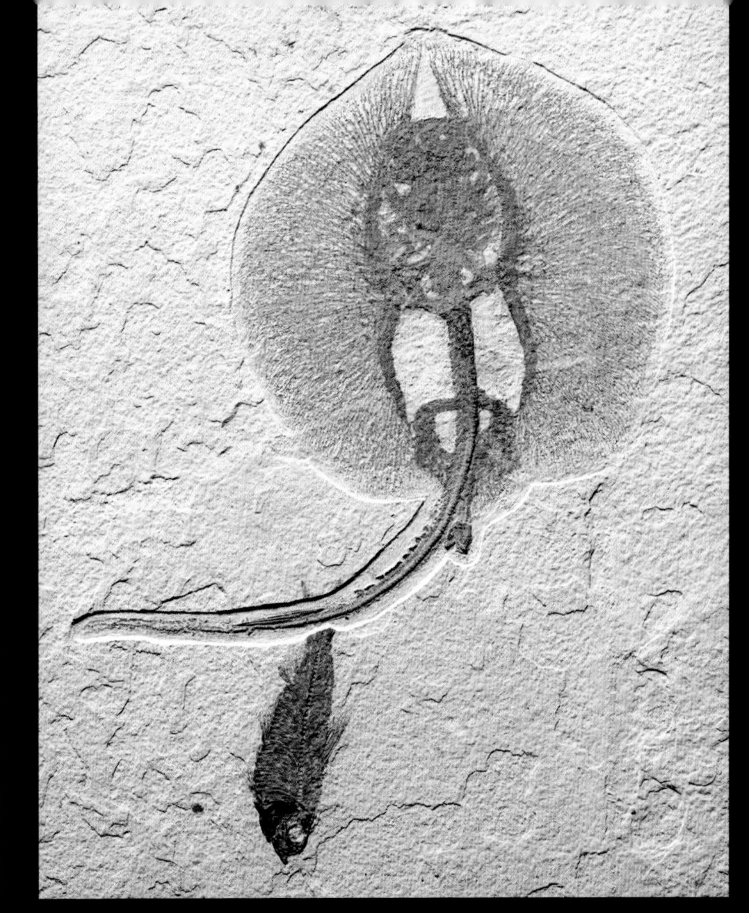

The oceans and waterways of earth abound with fish of many different types, such as the armored fish on the top left (Location: Quebec, Canada; Dated as Devonian), the shark on the bottom left (Location: Solnhofen Limestone; Dated as Jurassic), and the stingray (Location: Green River, Wyoming; Dated as Eocene). Each looks much like their modern-day counterparts. There are jawless fish, and fish with cartilage instead of bone. There are fish with lungs in addition to gills. Some look more like snakes, and some live in burrows. They inhabit many and varied habitats, from the deep ocean to coral reefs. But fish are fish, created to be fish on Day Five of the creation week. They did not descend from invertebrates, and did not evolve into land animals.

arranged in any order, an "evolutionary" story can be told about them. Try it with 1) a classic Coke® bottle, 2) an eight-ounce cardboard milk carton, and 3) a styrofoam coffee cup. Lined up in the order 1-2-3, we can tell a story that natural selection saw an evolutionary need for retaining heat, and thus enabled the provision of a less conductive outer coating. Lined up as 2-3-1, we could tell of the need for more height, so that the filling or emptying of the container would require less energy. In the order 3-1-2, we might envision the need for a square shape to provide more volume per inch of height. You can probably think up your own story to support the order 3-2-1. In reality, though, perhaps all three types appeared independently, at about the same time, and were designed by an intelligent designer to fill a specific need. A linear array (or a branching cladogram) tells us nothing of substance. Nor do the evolutionary "just so" stories so often told by evolutionists.

In a similar way, each of the the wide array of basic fish types seems to have appeared fully formed as that type, without a hint of relatedness or ancestral relationships. Stories can be told from the view of a particular researcher, but what do the data say? Recently, science writer A. N. Strahler wrote a lengthy anti-creationist book criticizing arguments from creationists. While enthusiastically embracing *Pikaia* as an ancestor for the vertebrates, he admits there is no record of the next step of diversification among the fishes.

> This is one count in the creationist's charge that can only evoke in unison from the paleontologists a plea of *nolo contendere*.[9]

Investigative Questions

Creationist Bill Jack has developed a simple way to help a student think through an issue and avoid being unnecessarily swayed by mere assertions masquerading as facts. He suggests a series of potent questions to ask of a claim, whether it is presented in a textbook, by an

rate, nor could the presentation of any be truly authoritative. Cladistics should certainly not be considered "proof" of evolution.

Evolutionists want everyone to be taught evolution and to believe that it relates an accurate portrayal of the history of life. They should be reminded that all competing cladograms assume evolution from the start and do not constitute evidence of, let alone proof for, the theory. Learners would be better off waiting for actual evidence to be discovered before committing to any view. Evolutionists should not expect students (or creationists) to choose between their competing opinions, for perhaps none of them is correct. Once a unified coherent idea is put forward, then it can be analyzed for accuracy and accepted or rejected. Until then, it is best to wait for pertinent evidence.

Of course, we have been waiting for factual confirmation for evolution for a long time. Creationists point out that the evidence that has actually been found directly fits their expectations.

Over the years, several fish candidates have been proposed as the ancestor of all land animals. None, however, has satisfied all researchers, and the controversy remains. Some champion the lobe-finned fish, which has bits of bone in its front fins. Others insist on the lungfish as the forerunner. Each school of thought has a preference, and in order to make its case, all other candidates must be discredited. These critiques are all convincing and expose serious weaknesses in each position.

Compare the bones in an amphibian forelimb or hind limb with the chips of bones in the crossopterygian's fin. While a rough correspondence can be made for the bones, great differences remain. The fin is laced with rays that give it strength, quite different from the phalanges in the hand or foot. Most importantly, the bones in the fin are not connected to the backbone, a condition that is necessary for walking.

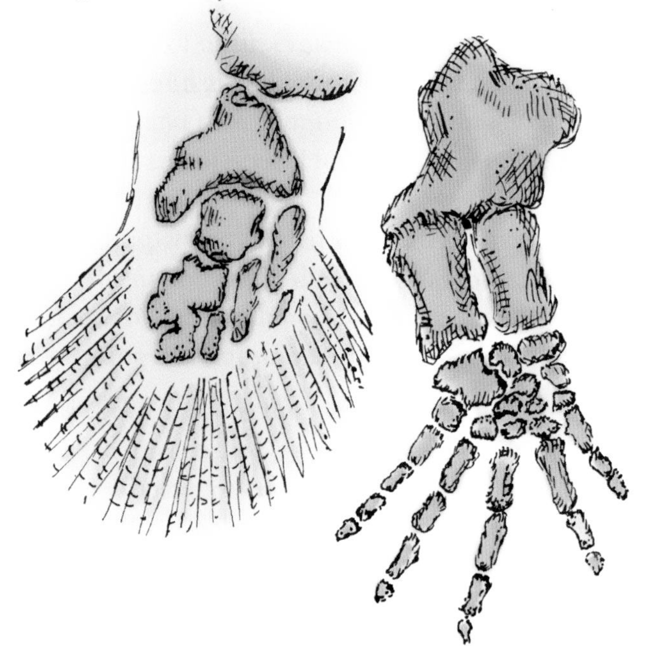

This may be due to the fact that the proponents follow essentially the same line of reasoning. First, identify a fish with some cartilage or bone in its fin. Next, show that the structure of this bone is similar to the vertebrate leg in a way that is superior to the structures present in other evolutionary contenders. Then, find other features—such as the pattern of the skull bones, teeth, vertebral column, or placement of the eye—and claim that this candidate has a closer relationship to some amphibian than all the other candidates have. Finally, a cladogram can be authoritatively drawn to show the proposed evolutionary lineage. Thus, the case is made.

None of the alternatives include a fossilized proto-leg, nor is there an evolving pelvic or shoulder girdle that can be exhibited. Likewise, hands and feet are completely missing from the discussions, because they are completely missing from all the fish fossils in question. What exists, instead, are fossils of different types of fish that possess none of the necessary transitional traits.

The best case for a fish becoming a four-legged amphibian would be made by exhibiting the evolving legs themselves, along with the newly evolved feet and hands. But in every case, the chips of cartilage or bone present in the fins are loosely embedded in muscle and not connected to the backbone at all. Lacking the necessary skeletal features, the fish are not even remotely equipped to stand and bear their weight.

The pseudo-bones in the fins are roughly analogous to the bones in the typical tetrapod's arm or leg, but they are never shaped, arranged, or employed in the same way that tetrapod bones are. There are sometimes extra bones, missing bones, or misplaced bones, but a rough correspondence can be made if the viewer has enough of a desire to see the link established.

Perhaps a better explanation for the arrangements of bone and cartilage would be a wise Creator who designed all creatures to fit the particular habitats of His choosing. The design that worked, altered to fit the specific need, was the one He employed. A variety of applications clustered around a working design—that is what we see.

Remember that similarity does not demonstrate ancestry. Each fish and each amphibian, reptile, mammal,

The coelacanth has long been considered a likely tetrapod ancestor. Many fossil specimens show chips of bone in the fins that were proposed as forerunners of leg bones. Thought to have been extinct for 100 million years, several communities of these unusual blue-green creatures were found alive in 1938 and in the years since. Close study of the living fish (species *Latimeria*) showed that it was quite different from what was expected of a transition. The ancestor of land animals is still debated among evolutionists.

and bird is seemingly well-designed to thrive in its environment. Each uses an amazing array of created applications of the basic bone arrangement common to all. The similarity better argues for a common designer than a common ancestor.

The lobe-finned fish was the most commonly accepted tetrapod ancestor for nearly a century, with speculation centering on the coelacanth. This broad fish type, which had been extinct—or so it was thought—for nearly 100 million years, exhibited chips of bone in its pectoral fin. Evolutionists speculated that it lived in a near-shore environment. Fins strong enough to battle waves and swim/crawl along the sandy bottom would theoretically have prepped it to walk ashore. Other opinions had this fish living in a shallow freshwater pond. When its home dried up, it then out-survived its contemporaries by "walking" to another pond.

But in 1938, a living lobe-finned fish—a coelacanth—was hauled up in fishing nets in the waters off Madagascar. Expectations were high, for now scientists could study a real missing link.

It has been estimated that at least 95 percent of an animal's uniqueness is found in its soft anatomy, its physiology, and not in its skeletal makeup. The recently discovered coelacanth (*Latimeria*) dashed evolutionary expectations as its heart, gut, and brain revealed characteristics that were vastly different from what had been anticipated. No feature was adapted to life on land, but was well-designed for life in the deep ocean. A similar evaluation awaited other lobe-finned fish and lungfish candidates.

Underwater studies eventually found that the strange fish lived in deep water. It did not dwell in shallow water, as the stories anticipated, acquiring the ability and equipment to go ashore. The strong fins were deftly used to make difficult swimming maneuvers at depth, not to support the fish's weight.

Fossils of the coelacanth fish have been found in rocks dated as old as 400 million years, yet the modern fish was essentially the same. Thus, no evolutionary changes—i.e., stasis—reigned for 400 million years. Meanwhile, the cousins of this fish supposedly evolved all the way to man.

Tiktaalik

The current darling of evolutionists is *Tiktaalik*, a recently discovered fossil fish with an unusual arrangement of bones in its robust pectoral, or front, fins. Several of these fossil fish have been discovered, each

Frogs found in the fossil record are recognizably frogs. They are not part frog and part non-frog ancestor. There are many different varieties of living frogs, and some extinct types. But variation and adaptation do not equal evolution. (Location: Germany. Dated as Eocene.)

Chapter 9 The First Land Animals

According to evolution, once animals had crawled out of the sea, they continued over many generations to adapt to life on land. Nearly every life function had to be changed. To begin with, air had to be utilized in a completely different way. Fish take in oxygen through gills. Lungfish can minimally survive prolonged time spans by breathing air (estivation), but their life cycle requires habitually breathing underwater through their gills.

Amphibians employ both gills and lungs at specific times during their lives, but this arrangement was not by chance. It was all in their original design. Did tadpoles acquire the genetic ability to breathe air through lungs by the mutation of genes that before then had aptly coded for gills? No, they already had all the necessary genetic information in their DNA at conception to use the appropriate organ at the appropriate stage of life. If an individual tadpole had to acquire the needed physiology to live on land, how could its gills continue to function while the new breathing apparatus was completed? Remember, breathing must be mastered at every stage of life without interruption.

Newborn amphibians breathe underwater, but they do not transform their gills as they grow. Instead, their entire body changes into something quite different. A tadpole does not look at all like a small frog. Although the change is gradual, it progresses along a predetermined path. This metamorphosis is not evolution, but simply growth over a frog's lifetime. At the tadpole stage, the frog is not yet fully grown, just as a human embryo is not yet an adult. They must have the genes already in place to provide the necessary functioning body parts at the right time.

The same could be said of legs and fins. Most fish cannot survive without fins of some sort, and most land-based amphibians need legs, which support their bodies' weight and enable them to stand. An "evolving" fish would need to acquire these or similar structures,

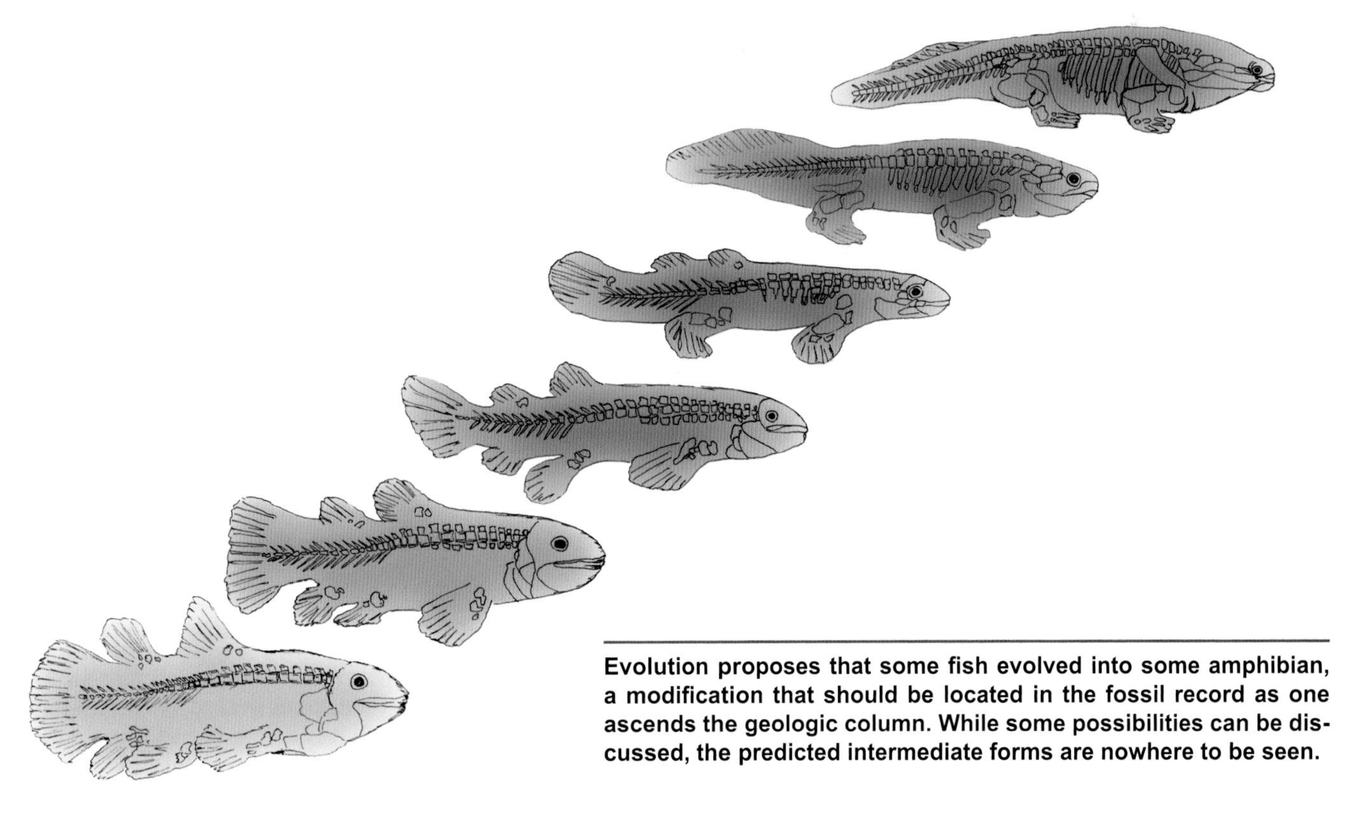

Evolution proposes that some fish evolved into some amphibian, a modification that should be located in the fossil record as one ascends the geologic column. While some possibilities can be discussed, the predicted intermediate forms are nowhere to be seen.

ent and do not cross. Might a more logical conclusion be that they were created separately by a wise and creative Creator for the functions they perform and the niches they fill?

Space does not permit a careful examination of each feature of these interesting and quite varied creatures. Both reptiles and amphibians flourish today in amazing variety, each diverse type in equilibrium within its environment, and exhibiting hardly a clue of common ancestry. We must, however, take the time to investigate some of the most striking and "popular" reptile types.

Marine Reptiles

> Understanding the 300-million-year old history of reptile life on earth has been complicated by...large gaps in the fossil record.[4]

Most reptiles habitually live on land. However, according to evolution, after leaving the sea and acquiring all the necessary physiological and skeletal features to dwell on land, some reptiles ventured back into the water to live. These included the ancestors of the ichthyosaurs, shark-like creatures that were highly specialized for life in the deep. What conditions would cause a successful land-dweller to return to an environment from which a lengthy series of random, but "beneficial," mutations had extricated it?

Evolutionists must employ a kind of reverse Darwinian psychology here. Darwin's followers argue for "survival of the fittest" in the assumed transition from water to land. They claim that while new and someday useful traits were being acquired through mutations (such as legs from fins), the "misfit mutants" were able to survive due to a lack of competition in the new environment. That "just so" story will not work, however, in the transition from the land back into the sea, since those waters were teeming with myriads of well-designed competitors.

Evolutionists believe that once ichthyosaurs got to the water, they functioned quite efficiently, aptly filling an important ecological niche. Were conditions in the sea somehow more advantageous to land-dwelling reptiles? According to the evolutionary story, after eons of trial and error al-

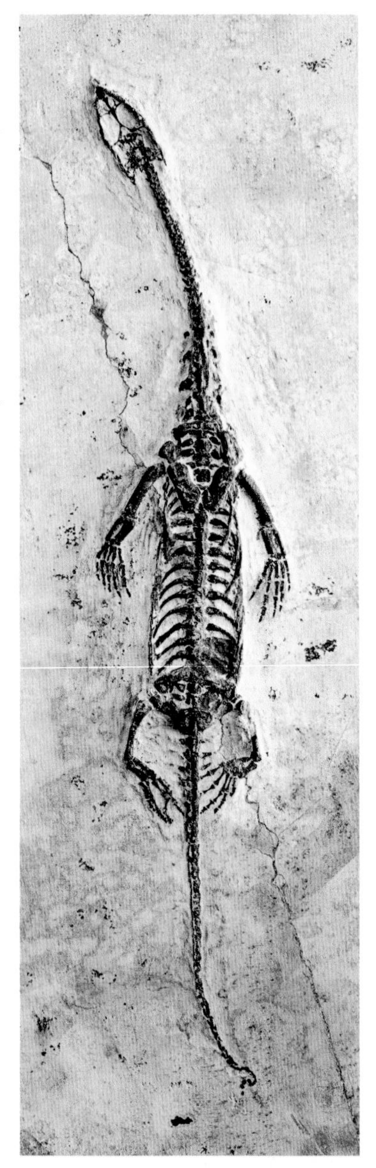

Swimming reptiles like the plesiosaur, ichthyosaur, and this aquatic reptile (right) were quite specialized, having been equipped to thrive in hostile waters. (Location: China. Dated as Triassic.)

Science learned from this fossil that this reptile gave birth to live young, rather than laying eggs. The ichthyosaur must have been buried extremely quickly, for the entire process of giving birth would not have taken a long period of time. This is not a record of the slow accumulation of sediments around a dead carcass, but a record of death by rapid burial, as sediments cascaded around the struggling animal.

terations, the new marine reptiles eventually became skilled hunters that were unable to leave the water, even for reproduction. But why did they go there in the first place? All that is known for sure is that they thrived for a while and then eventually all went extinct.

One might think evolution would propose that sharks gave rise to the shark-like ichthyosaurs, and then they to the land-based reptiles. But this is not so. The timing of their fossil appearance negates that possibility. If the dating is correct (which we might dispute), land reptiles came before marine reptiles.

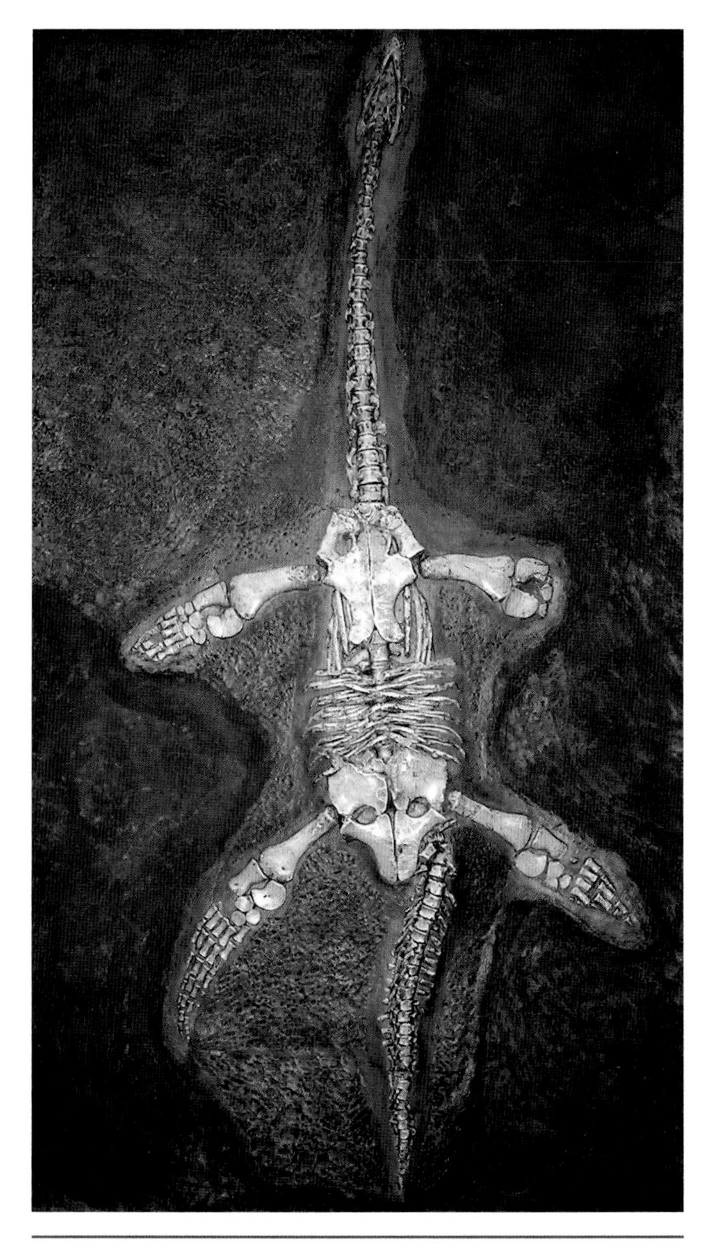

The well-known plesiosaur is thought by evolutionists to have evolved from land reptiles. They are so specialized for life in the water, however, that one wonders how random mutations could have accomplished this. Note the front and rear "feet" of this specimen. They are not at all like the feet of other reptiles, yet they are necessary for survival in the water. They have the same sorts of bones, but are quite different in their application. Proposed transitional creatures between the two groups are unknown. Evolution's "story" of transition lacks credibility.

Ichthyosaurs supposedly retained their recently acquired lungs, but they abandoned their forelegs and hind legs in favor of equally complex and specialized paddles—all without leaving a hint in the fossil record of the transition. They had a robust tail, much like huge fish, that was capable of propelling the creature with great force and agility through the water. Some female fossil ichthyosaurs have been found preserved in the process of giving live birth. This not only demonstrates the rapid burial and preservation of the ichthyosaurs' remains, it also reveals a reproductive strategy nearly unique among reptiles, and one that was not employed by any proposed ancestor. Ichthyosaurs appear abruptly, fully formed and functional, without the necessary ancestral linkage. No evidence has been touted that they evolved from any other reptile. And numerous fossils of icthyosaurs have been found, so there can be no appeal to the paucity of the record.

> The basic problem of ichthyosaur relationships is that no conclusive evidence can be found linking these reptiles with any other reptilian order.[5]

Much the same could be said of the plesiosaurs, giant marine reptiles that are often linked with the dinosaurs in popular treatments. With very large paddles and long necks, they have little similarity to other reptiles, but they too are supposed to have descended from land reptiles. Unfortunately, no ancestral connection with any possible non-marine reptile has been found. They stand alone, separate and distinct from all other reptiles, having appeared abruptly, already exhibiting design precision, and equipped to do what they did very well. These giant examples of creation may have gone extinct, but their testimonies remain.

Flying Reptiles

Reptiles that flew (pterodactyloids and rhamphorhynchoids) form a group that was distinct from all other reptiles. The two suborders are themselves quite different from each other, offering no clue as to their relationship. The pterosaurs are known for a long, bony crest on top of the skull, while the rhamphorhynchoids commonly sported a heavy knob on the end of their long tail. Fossils with wingspans up to 54 feet have been discovered. How could they fly?

And from what could they have evolved? Evolutionists surmise they must have descended from a walking di-

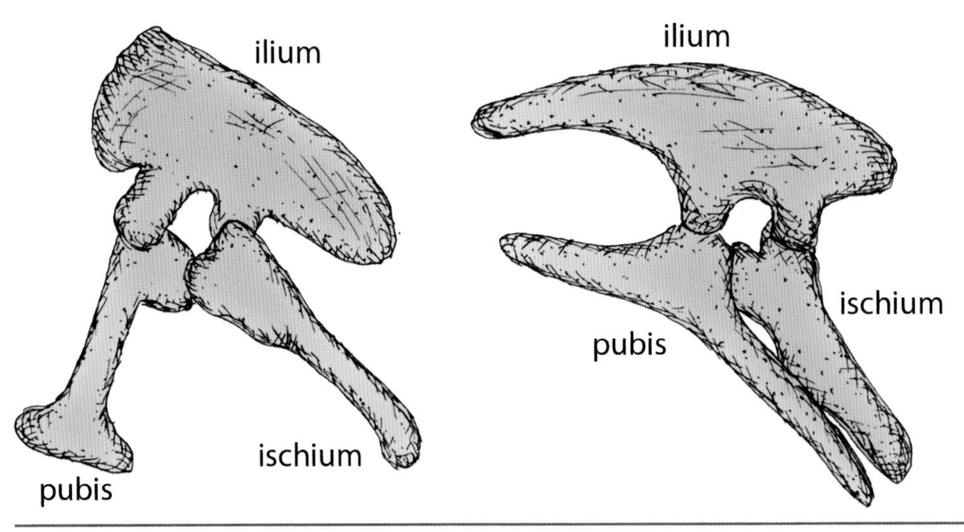

Reconstructions of the two types of hip possessed by all dinosaurs. No modern reptile has either hip. Those that walk today have the standard reptile hip, with a sprawling gait. Compare both kinds with the radically different bird hip shown on page 77. The popular claim that birds evolved from dinosaurs rings hollow when the specifics are considered.

***Edmontosaurus*, a member of the Hadrosauridae family, was part of a well-known group of "duck-billed" dinosaurs that are easily recognizable by their distinctively flattened, duck-like beaks. They were so numerous and traveled in such large herds that they have been referred to as the "cattle of the Cretaceous." They were most likely the favorite prey of *Tyrannosaurus rex*, as evidenced by the numerous *Hadrosaurus* bones that have been discovered with *T. rex* teeth marks.**

birds. Museums and textbooks unabashedly parade this disputed relationship before an unsuspecting public, omitting the many inconsistencies pointed out by scholars, even in the important hip structure. Not only do birds not possess a "bird hip" or "lizard hip" as dinosaurs do, they do not even use their hip when they walk. In walking, birds use only the bones from the "knee" down. Virtually all tetrapods walk by moving the thigh bone, including those dinosaurs from which birds are thought to have evolved. This strongly implies that dino-to-bird advocates are mistaken.[10]

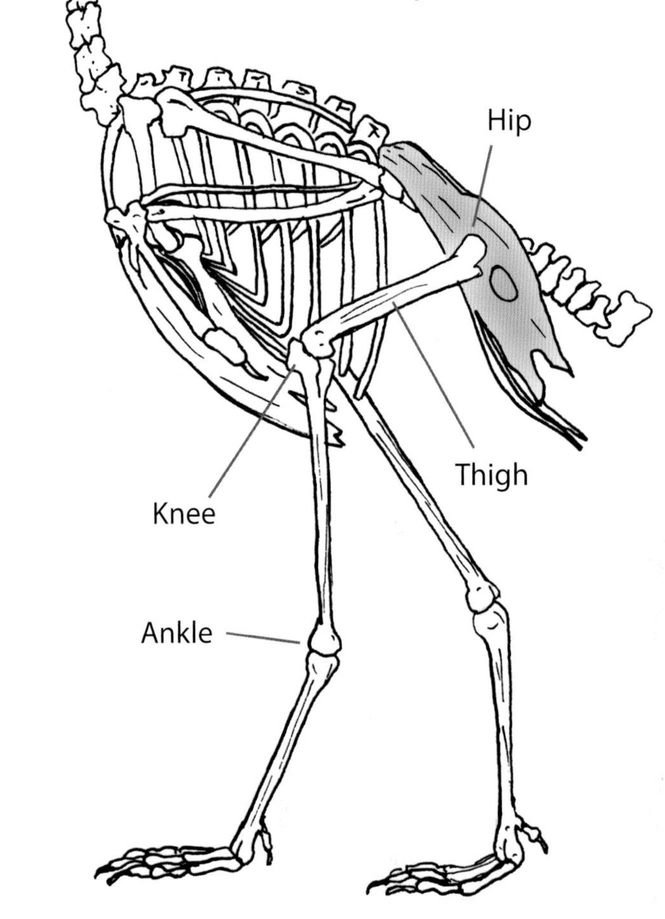

A bird's thigh does not move substantially from its nearly horizontal position, where it provides rigid lateral support to the thin-walled air sacs of the respiratory system. Birds are "knee walkers," not hip walkers. This poses a seemingly insurmountable barrier to the dino-to-bird theory.

And a bird's high metabolism, which is quite different from that of cold-blooded reptiles, requires about 20 times more oxygen than reptiles. This is made possible in part by their stationary thigh bone, or femur. Connected to the rib cage, it supports the diaphragm and lung system. This allows air (and the needed oxygen) to be drawn in nearly continuously, without being interrupted by breathing in and out. It is a unique setup in the animal kingdom, but without it birds could not take in enough oxygen or support their high metabolism and blood circulation rate.

The fact that dinosaur fossils have been found from pole to pole can partially be explained by plate tectonics moving the continents around. Without a doubt, the earth's surface today is divided into plates, and much evidence exists that they separated subsequent to the deposition of their sediment layers. There are many weaknesses in the standard plate tectonics concept, most particularly the lack of an adequate mechanism for slowly moving such massive plates. Probably the best reconstruction of past continental movements can be found in the paper "Catastrophic Plate Tectonics: A Global Flood Model of Earth History."[11] This innovative concept calls for rapid plate-spreading not very long ago, due to tectonic forces acting at rates, scales, and intensities that far exceeded their counterparts operating today. Rapid subduction of the separating plates dragged the oceanic lithosphere far down into the earth's interior—where their remnants currently reside near the core-mantle boundary, as located by seismic studies—and opened new oceans. In this model, the separating surface plates seem to have almost come to a halt in modern times. Thus, the surface continents are not now in the locations where they originated. This helps explain why some fossils are located in strange places.

The fossils and their related sediments have long testified that a relatively warm climate existed over nearly the entire globe throughout much of the past. In this environment, the dinosaurs thrived. Most dinosaurs were herbivorous, and many plants in the past were larger than their modern counterparts, often growing in lush subtropical rainforests. All environments past or present must contain many more plants than planteaters to support the food chain, and the dinosaurs were big eaters. Even though the total volume of dinosaurs was impressive, the plant volume far surpassed it.

Evidence of this plant volume can be found in the fossil record. Some evolutionary histories of the earth wrongly portray dinosaur death and burial as producing the world's oil reserves. Coal is metamorphosed plant material, and while natural gas can come from a multitude of sources— both inorganic and organic, including the rotting of animal carcasses—liquid hydrocarbons

ed sunlight from reaching earth, interrupting plant growth and the animal food supply. Thus, no dinosaur survived beyond the beginning of the Tertiary system. This is known as the K-T boundary. But several dinosaur remains have been found in Tertiary rocks after the K-T boundary, thus challenging the asteroid impact hypothesis.[15]

The imagination can run wild when only meager evidence is available for support. Whatever the cause for dinosaur extinction, it involved unimaginable forces operating on a vast scale. The sediments in which the dinosaurs were typically deposited represented a watery flow of terrestrial mud that devastated large portions of the globe. Whether or not it was triggered by an asteroid, that sounds like a great flood just like the Flood described in the Bible.

Soft Dinosaur Tissue

A recent discovery has opened up entire new areas of study.[16] A large tyrannosaur thigh bone, deemed to be 70 million years old, had to be broken in two for transportation. Surprisingly, inside the bone Dr. Mary Schweitzer's team found soft tissue that was flexible and pliable. Its hollow blood vessels contained what looked like actual blood cells. This shocked researchers. Even under ideal circumstances, organic tissue like this is known to break down in a rather short period of time, especially in the presence of water, and these conditions were far from ideal. Even fragile proteins were present, though partly degraded, indicating that this specimen must have been fairly recently deposited—far more recently than its assigned age.[17] Soft tissue, DNA, and blood cells—as well as "fresh-looking bone"—had previously been discovered in a number of "old" fossils from a variety of rock

This nest of eight dinosaur eggs has tentatively been attributed to the strange-looking theropod dinosaur *Therizinosaurus* (previously called *Segnosaurus*). Little is known about this elusive beast. No skull has ever been recovered. The first bones to be discovered were thought to be from a turtle-like reptile, hence the species name *cheloniformis*, which means "turtle-formed." Nonetheless, they are believed to have grown up to 32 feet in length, and can be characterized by their long necks and Ornithischia-like skeletal anatomy. Notable characteristics include partially fused, backward-facing hip bones, and their forelimbs boasted extraordinarily long claws, which grew up to three feet in length. These fearsome weapons would have been invaluable in defense. (Location: Gobi Desert. Dated as Cretaceous.)

Soft and stretchy blood vessels with blood cells found inside a dinosaur bone give evidence of a much more recent burial than the seventy-million-year age normally ascribed to it. All studies conclude that such tissues could not survive long ages, yet here they are. Could the dating technique be in error? These photographs and their captions make a strong point in favor of a young-earth interpretation.

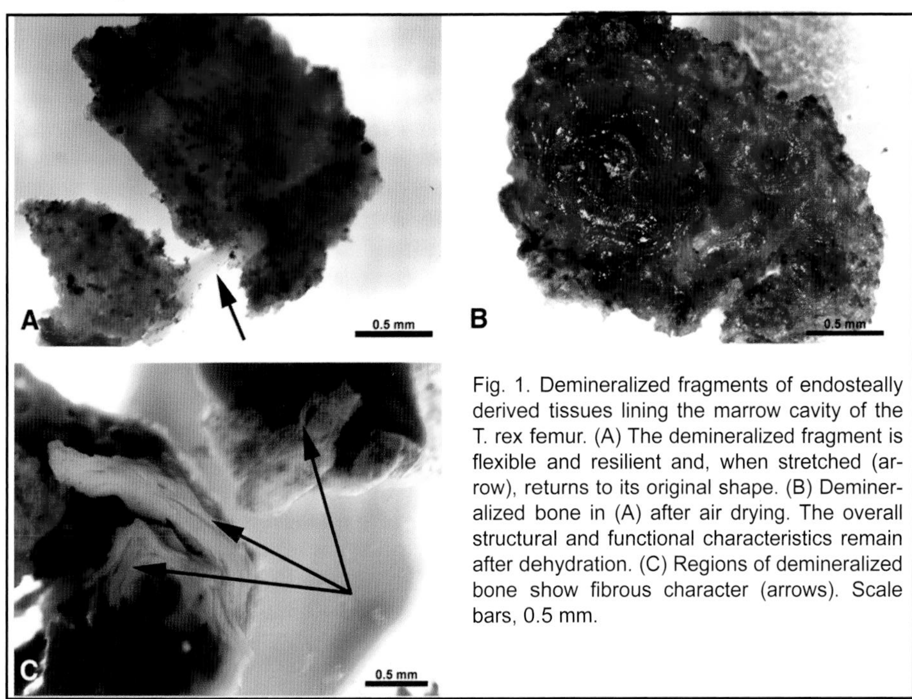

Fig. 1. Demineralized fragments of endosteally derived tissues lining the marrow cavity of the T. rex femur. (A) The demineralized fragment is flexible and resilient and, when stretched (arrow), returns to its original shape. (B) Demineralized bone in (A) after air drying. The overall structural and functional characteristics remain after dehydration. (C) Regions of demineralized bone show fibrous character (arrows). Scale bars, 0.5 mm.

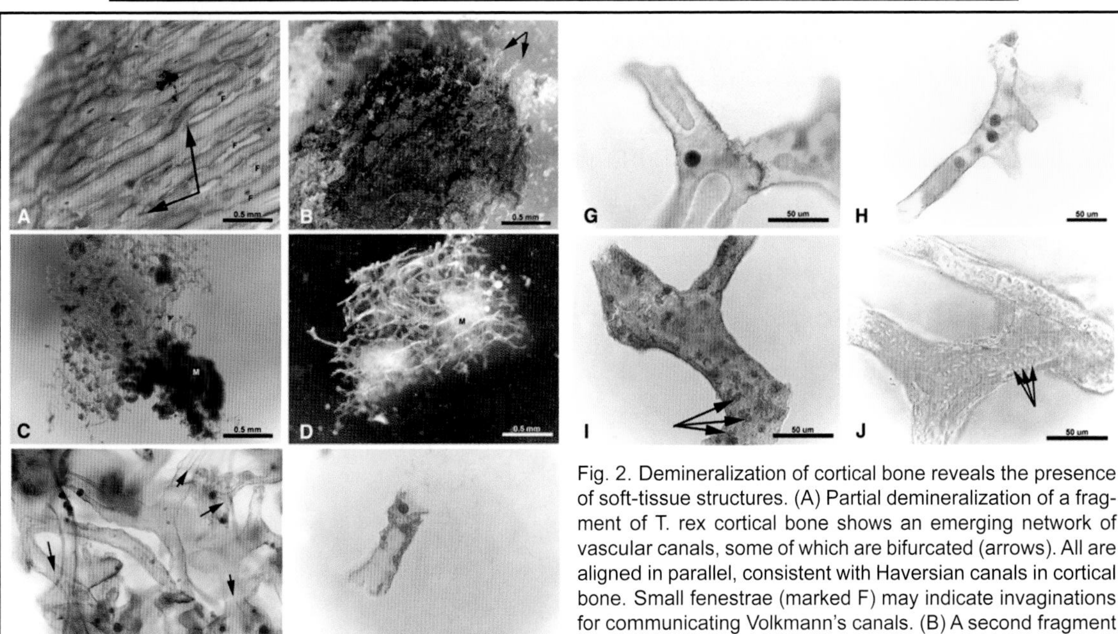

Fig. 2. Demineralization of cortical bone reveals the presence of soft-tissue structures. (A) Partial demineralization of a fragment of T. rex cortical bone shows an emerging network of vascular canals, some of which are bifurcated (arrows). All are aligned in parallel, consistent with Haversian canals in cortical bone. Small fenestrae (marked F) may indicate invaginations for communicating Volkmann's canals. (B) A second fragment of T. rex cortical bone illustrates transparent vessels (arrows) arising from bone matrix in solution. (C) Complete demineralization reveals transparent flexible vessels in what remains of the cortical bone matrix, represented by a brown amorphous substance (marked M). (D) Ostrich vessel after demineralization of cortical bone and subsequent digestion of fibrous collagenous matrix. Transparent vessels branch and remain associated with small regions of undigested bone matrix, seen here as amorphous, white fibrous material (marked M). Scale bars in (A) to (D), 0.5 mm. (E) Higher magnification of dinosaur vessels shows branching pattern (arrows) and internal contents. Vascular structure is not consistent with fungal hyphae (no septae, and branching pattern is not consistent with fungal morphology) or plant (no cell walls visible, and again branching pattern is not consistent). Round red microstructures within the vessels are clearly visible. (F) T. rex vessel fragment, containing microstructures consistent in size and shape with those seen in the ostrich vessel in (H). (G) Second fragment of dinosaur vessel. Air/fluid interfaces, represented by dark menisci, illustrate the hollow nature of vessels. Microstructure is visible within the vessel. (H) Ostrich vessel digested from demineralized cortical bone. Red blood cells can be seen inside the branching vessel. (I) T. rex vessel fragment showing detail of branching pattern and structures morphologically consistent with endothelial cell nuclei (arrows) in vessel wall. (J) Ostrich blood vessel liberated from demineralized bone after treatment with collagenase shows branching pattern and clearly visible endothelial nuclei. Scale bars in (E) to (J), 50 μm. (F), (I), and (J) were subjected to aldehyde fixation (3). The remaining vessels are unfixed.

From M. H. Schweitzer, J. L. Wittmeyer, J. R. Horner, and J. K. Toporski, 2005, Soft-Tissue Vessels and Cellular Preservation in *Tyrannosaurus rex, Science*, 307 (5717): 1952-1955. Reprinted with permission (and at Dr. Schweitzer's insistence without changes) from AAAS.

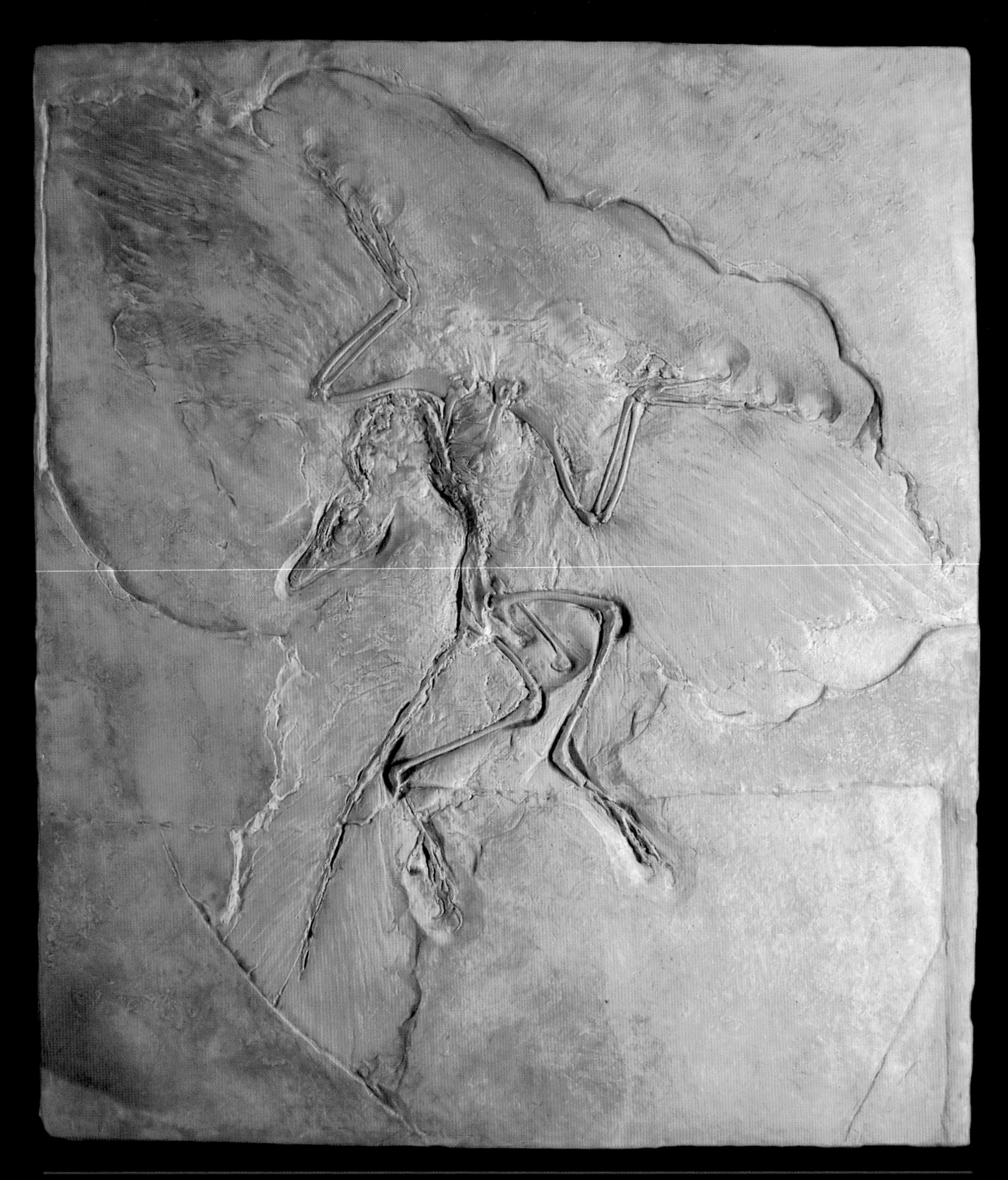

Archaeopteryx possessed features in common with both birds and reptiles, as well as features found in all vertebrates. The diagnostic traits, however, were those of a modern bird, from a perching foot to a robust sternum. This bird exhibited a mosaic of traits, not a transition between reptiles and birds. Furthermore, the date assigned to these fossils is far earlier than would be expected for this supposed transition.

Chapter 10 Fossils of Special Interest

When Darwin wrote his famous book *On the Origin of Species*, he was worried about the lack of transitional fossils and had high hopes they would one day be found.

> Why, if species have descended from other species by insensibly fine gradations, do we not everywhere see innumerable transitional forms? Why is not all nature in confusion instead of the species being, as we see them, well defined?... But, as by this theory innumerable transitional forms must have existed, why do we not find them embedded in countless numbers in the crust of the earth?...Why then is not every geological formation and every stratum full of such intermediate links? Geology assuredly does not reveal any such finely graduated organic chain; and this, perhaps, is the most obvious and gravest objection which can be urged against my theory.[1]

Despite the lack of fossil support for the evolution of life, many people—especially the religious and political leaders of his day—quickly embraced Darwin's new theory and shared his frustration in the lack of positive fossil evidence. A concerted effort ensued that sought to find a solution to this "most obvious and gravest objection."

Imagine their and Darwin's elation when a striking fossil was found soon thereafter in the Solnhofen Limestone in Bavaria, which has a fine-grained, lithographic-quality matrix. Named *Archaeopteryx*, the fossil was hailed (and still is, in many circles) as the perfect example of a transitional form between reptiles and birds. It had several reptile-like characteristics: teeth, claws on its wing, and a tail. It also possessed various bird-like characteristics, such as fully formed feathered wings, a perching foot, and a wishbone. It seems like a perfect missing link—or is it?

As it turns out, none of these traits is unique to either reptiles or birds. No definitive statement can be made as to *Archaeopteryx's* place in the evolutionary tree just on the basis of these shared traits. For instance, while no living birds have teeth, some fossil birds do have teeth, so this trait is hardly a defining characteristic. Furthermore, some modern birds have wing claws, so to find a fossil bird with wing claws says little about whether it came from a reptile. *Archaeopteryx* contains an unusual combination of functional traits—Stephen Gould called it a "mosaic" of traits[2]—but no transitional ones. Nevertheless, even though its size is roughly that of a pigeon, *Archaeopteryx* has been elevated to "rock star" status, and the half dozen or so specimens discovered (two fairly complete) get top billing in books on evolution.

Despite its popular reputation, however, *Archaeopteryx* no longer holds such a position among knowledgeable evolutionists. Many have noted that its features are not what would be expected of the ancestor of all birds. It actually causes more problems than it solves. For one thing, by any current definition it is a modern bird, with a complete wing and fully modern feathers. It also has a perching foot and a robust wishbone, just right for a flying bird.

Even if it were a bird/reptile, its place in the evolutionary tree is far from resolved. The date assigned for its earliest representative is problematic.

> The earliest undisputed bird fossil is *Archaeopteryx*, found in the upper Jurassic (145 million years ago). There are two major proposals concerning the phylogeny of birds. According to the thecodont theory, birds originated from archosaurian reptiles more than 200 million years ago [i.e., birds evolved before *Archaeopteryx*]. According to the dinosaur theory, birds originated from theropod dinosaurs in the later Cretaceous (ca. 80-100 million years ago) [i.e., the origin of birds came long after *Archaeopteryx*, itself a true bird].[3]

Either way, *Archaeopteryx* becomes irrelevant to evolution's timeline for the appearance of birds. Furthermore, at least one of the *Archaeopteryx* specimens possessed a bony sternum where the ribs met in front, which is needed as an anchor for the powerful muscles required for flight. However, few reptiles had ribs that even covered the front.

And so, *Archaeopteryx* does not fit well with any evolution theory. The dating of relevant fossils strongly testifies against the idea that *Archaeopteryx* (usually assigned to the Late Jurassic, about 150 million years ago) and other birds descended from dinosaurs. The most probable ancestor for *Archaeopteryx* is dated at about 125 million years old. How could the descendant be older than the ancestor? Cladistics theory, now popularly used to construct hypothetical family trees of the past, necessitates that the few questionable fossils that have been found are dominated in importance by theoretical creatures that have never been found. Evolution drives the bus, while the facts take a back seat.

Paleontologists have recently discovered evidence of birds' existence that long predates *Archaeopteryx*, according to standard dating. There have been fossil tracks of birds found in strata dated at 212 million years ago (in the Late Triassic), long before *Archaeopteryx*.[4] And controversial fossils of a modern-appearing bird named *Protoavis* were also found in strata of Late Triassic age.[5]

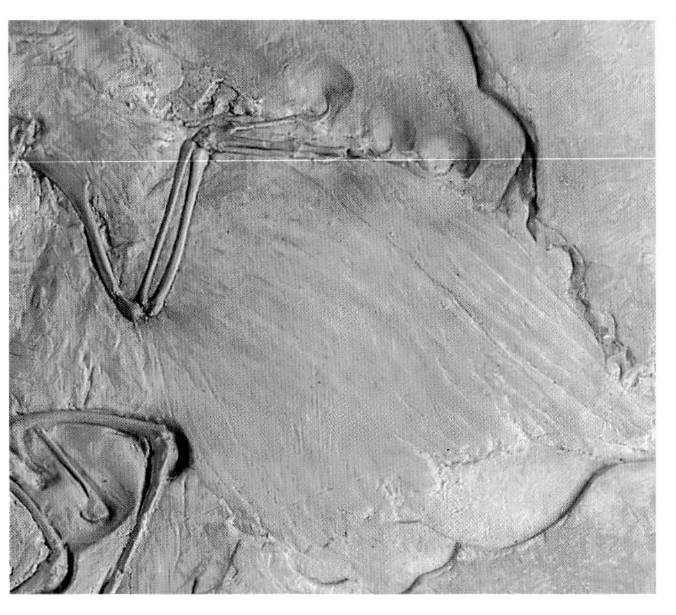

Close-up of *Archaeopteryx* wings. Asymmetric feathers can be seen, with shaft, barbs, and barbules, just like feathers on modern flying birds. This was a flying bird.

It is not enough that *Archaeopteryx* has some skeletal features in common with reptiles, for so does every bird or mammal alive today. Does this imply relatedness? How about its soft anatomy? Does *Archaeopteryx* have the necessary physiology as well as skeletal features? In considering what would be necessary to alter a reptile into a bird, the following abridged list of evolutionary obstacles might be helpful.

Wings: The proposed ancestors of birds are controversial. Some think they must have been bird-hipped dinosaurs with two diminutive front legs. These supposedly gained the ability to fly by running and hopping for insects while beneficial mutations accumulating over vast ages allowed their front limbs to lengthen and become wings. The tiny forelimbs of these dinosaurs had digits similar to a hand, consisting of digits one, two, and three. But bird forelimbs have digits two, three, and four.

Other theorists rely on cladistic analysis to conclude that bird ancestors were lizard-hipped theropods. They hold that ground-dwelling dinosaurs climbed and jumped out of trees while their wings, fraying due to many successive mutations, eventually enabled them to fly. In either case, flight requires fully formed, interlocking feathers and hollow bones, not to mention flight muscles anchored to a sternum.

Feathers: The feathers of flying birds require precise design, and are not at all similar to reptile scales. Even if scales became "frayed," the shreds would not be interlocking and effectively impervious to air, as are feathers. Actually, the shafts of feathers are more similar to hair follicles on mammalian skin than they are to scales on a reptile. Could such precise feather design on birds arise by mutations of the horny-textured plates on reptiles? In all the recent discoveries of dinosaur fossils with "feathers," the controversial feathers are merely inferred. Many experts doubt the feather interpretation, insisting that the impressions are better understood as reptilian frills or as thin filaments of decomposing tissue that originated under the skin.

> Our findings show no evidence for the existence of protofeathers....We conclude that "protofeathers" are probably the remains of collagenous fiber "meshworks" that reinforced dinosaur integument [i.e., surface layer]. These "meshworks" of the skin frequently formed aberrant patterns resembling feathers as a consequence of decomposition.[6]

Bones: Birds have delicate, hollow bones designed to lighten their weight, while most dinosaurs had solid bones. Dinosaur bones that were hollow usually included the skull and other bones of the giant dinosaurs, such as land-confined sauropods, which are not considered candidates for bird ancestors. The placement and design of bird bones might be somewhat analogous to those in dinosaurs (as they are in all vertebrates), but they are actually quite different. For example, the heavy tail of dinosaurs (needed for balancing on two legs) would prohibit possible flight. Besides, the theropods (like *T. rex*) that are proposed as possible bird ancestors were "lizard-hipped" dinosaurs, not "bird-hipped," as might be expected.

Warm-blooded: Birds are warm-blooded, with exceptionally high metabolisms and food demands. While dinosaur metabolism is in question, all modern reptiles are cold-blooded, with a more lethargic lifestyle. It is very difficult to ascertain from only skeletal material whether dinosaurs were warm-blooded or cold-blooded, and each position has many proponents. To accommodate all the data and speculations, some now hold that dinosaurs were warm-blooded and grew quickly while they were young, but altered their metabolism as they aged and became cold-blooded and slow-growing. That, however, does nothing to answer the question of how warm-bloodedness—a complex, multifaceted physiological adaptation—originated in the first place.

Lungs: Bird lungs are supported by a stationary hip and femur. Being "knee walkers" allows birds to breathe quite differently from reptiles and mammals. In birds, air flows continually in a one-directional loop through the lungs. Reptilian respiration is entirely different, being somewhat similar to the "in and out" breathing of mammals. The ability to breathe must function at every moment and in every individual creature. In order to be selectable, slowly accumulating mutations for intermediate body parts must work and be advantageous at every stage of transition. Switching breathing styles would have to be instantaneous and functional, or death and extinction would result.

Other Organs: The soft parts of birds and dinosaurs, in addition to the lungs, are totally different. A recently discovered "mummified" dinosaur with fossilized soft tissue proved to have a heart quite like a reptilian crocodile and not at all like a bird. Each generation needs a beating heart.

The Origin of Flight

The marvels of bird flight seemingly testify to intelligent aerodynamic engineering. The interlocking feathers, wings, hollow bones, sternum, flow-through lung, and flight muscles are all designed specifically for flying. This suite of features is only useful for flight, and they all are necessary for any part to accomplish its purpose. Random mutation and natural selection would have been hard-pressed to accomplish this.

But according to evolution, flight evolved on several occasions. Happening once is highly unlikely. Happening multiple times—how can evolutionists assert this and maintain a straight face?

Flight was "first" achieved by insects. The insects are classed as arthropods, as are many invertebrate animals. Only some of them took to the air. Yet the wide variety of flying insects astounds those who try to catalog them. From the delicate butterfly to the aggressive dragonfly to the communal honeybee to the filthy housefly to the irritating gnat, all exhibit precise design characteristics and yet bear little evidence of relatedness. Both living specimens and fossil specimens are easily identified and are seemingly designed to do what they do. There are neither transitional forms among them nor linking them with non-flying insects. They testify to purposeful creation, not random evolution.

The flying reptiles are likewise separate and distinct from all other reptiles. From the lowest strata in which their fossils are found, they display all the design traits that characterize them. The two types of flying reptiles are quite different from each other and lack fossil evidence that they evolved from some other type. They seem to have been designed to be flying reptiles only, created precisely with that goal in mind.

Birds are not thought to have evolved from flying reptiles, but from ground-dwelling or tree-climbing reptiles. Birds fill diverse ecological niches and accomplish numerous necessary purposes, all the while filling the air with song and beauty.

Some mammals, too, can fly. Bats exhibit many unique design features, including wing design and sonar. They accomplish several necessary tasks, including keeping insect populations in check, without which our lives would be difficult. Bat fossils are 100 percent bat. No evolution here!

Thus, the dinosaur-to-bird transition is blocked by many major obstacles, not just the acquisition of feathers. To make the situation even worse, in order to make this transition, most if not all of the definitive characteristics must be acquired simultaneously. All of the intermediate parts must be present and working or else none serves a valid function that could be selected. Evolutionary explanations do not fit the facts. Stories should not be allowed to substitute for data.

The Death Pose

Dinosaur fossils are often discovered encased in water-deposited sediments, with their heads thrown back, their tails curled forward, and their mouths wide open. It has been speculated that as their bodies dried out, the muscles and tendons contracted to produce this pose. But remember, animals must be quickly buried or they will be eaten by scavengers or completely decompose and not be fossilized at all. These fossilized creatures must have been buried as they choked on sediments, much as would happen in quicksand.[7]

Note the pose of *Archaeopteryx* fossils depicted in popular photos. Virtually all *Archaeopteryx* fossils show the bird with its head arched back, as if gasping for air. Paleontologists have traditionally speculated that rigor mortis after death arched the back, but birds do not normally assume such a pose when they die. Dare we evoke a rare exceptional circumstance as the rule for *Archaeopteryx*?

These birds seemed to have been trapped in rapidly moving water. They evidently died a horrible death by drowning, vainly trying to draw a breath while being surrounded and covered by rapidly accumulating sediment. Even birds that die while flying over water would not drown like this. Some birds can swim, but *Archaeopteryx* was neither equipped for aquatic life nor showed signs of having died naturally. If it had, the carcass would not have retained its death pose, nor would it have likely remained intact. Bird carcasses are rather fragile and rapidly disintegrate as their bones scatter. These birds suffocated in an unnatural watery environment that was continually receiving sediments. They were buried alive and preserved, their last pose gasping for breath serving as a testimony to the conditions of their deaths.

This violent death pose extends throughout the fossil record. For instance, clams are able to burrow through sediment, but clam fossils are often found "clammed up," with both halves tightly shut. This is a living posture for a clam that has been disturbed and feels in danger. It appears they were quickly and catastrophically buried, encased in sediment too deep for them to burrow out.

Fish or other marine fossils are sometimes found "in action poses" that were dramatically interrupted, sometimes in the process of eating another creature, thrashing about, or while giving birth. They died suddenly and violently, being quickly overwhelmed by choking sediments that buried and fossilized them.

Sue

Another recent discovery has opened a window into the days of the dinosaurs, telling an eloquent story about the circumstances in which they lived, as well as how they died. In 1990, a nearly complete *Tyrannosaurus rex* fossil was discovered in the badlands of South Dakota. Named after someone involved in the discovery, "Sue" stood over thirteen feet tall at the shoulder and measured 42 feet from head to tail. She had frightening teeth, some with serrated edges. Evidence of a carnivo-

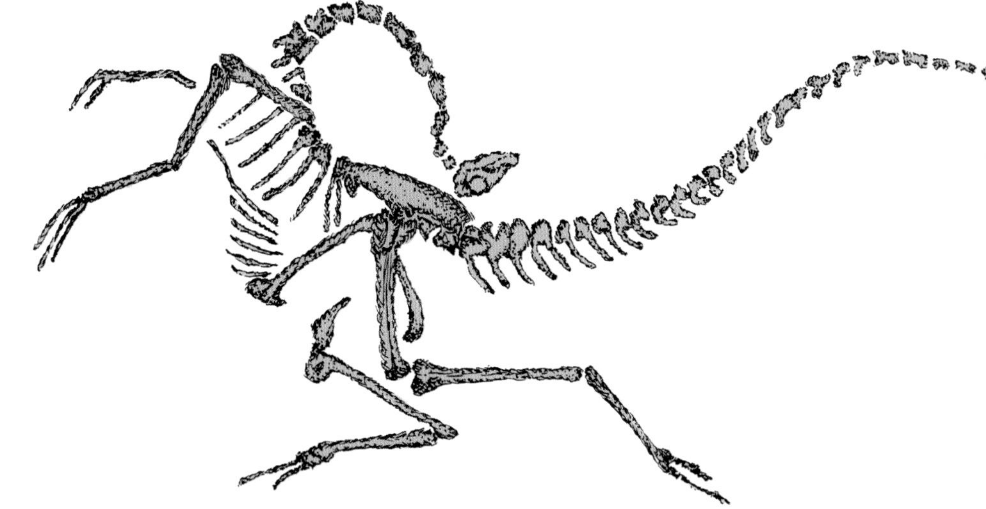

Substantially all *Archaeopteryx* fossils and many dinosaur fossils worldwide are found seemingly caught in the throes of death, evidently asphyxiated. Were they all drowned? Perhaps a worldwide Flood was responsible. This dinosaur is *Struthiomimus*, a bird-hipped, ostrich-like specimen with a foot-long skull found in 1914 in Canada. (Dated as Cretaceous.) Redrawn from Faux and Padian, 2007.

"Sue" is perhaps the largest and most complete *T. rex* ever found. The skeleton speaks of great violence surrounding its life and death. Note the teeth marks in its jaw, matching the teeth of another *T. rex*. The puncture wounds healed, but most likely had caused painful abscesses. The femur was broken. It had partially reset, but could bear little weight. Several ribs had also been broken and reset. No doubt Sue experienced an existence of horrible pain, followed by a death of violence in the great Flood of Noah's day.

rous diet can be seen in other *T. rex* fossils in their partially digested stomach contents and in fossilized dung remains.

Sue's life history, however, is written in her bones. They indicate, among other things, that she lived for quite some time by modern standards and led a very violent life. She bore lasting scars from several encounters with combatants. Puncture wounds in her jaw match the tooth pattern of another *T. rex*. These wounds abscessed and then healed. She also had several broken and partially healed ribs and a broken left leg that had "set" itself incompletely. These injuries did not kill her, but they must have made the remainder of her life miserable. How could she hunt and defend herself from other hunters? Her life must have been unspeakably vicious and painful.

Many have noticed similar scars on nearly all dinosaur fossils. Evidently, the time of the dinosaurs was exceptionally violent and bloody. Evolutionists designate this period as the "Age of Dinosaurs," while creationists assign it to the days immediately prior to the great Flood of Noah's day. All things lived in a violent world of predators, prey, and scavengers, disease and injury, in which hunters sometimes became the hunted.

Of course, this is exactly how the Bible describes the days before the great Flood of Noah's day.

> The earth also was corrupt before God, and the earth was filled with violence. And God looked upon the earth, and, behold, it was corrupt; for all flesh had corrupted his way upon the earth. And God said unto Noah, The end of all flesh has come before me; for the earth is filled with violence through them; and behold, I will destroy them with the earth. (Genesis 6:11-13)

Did the Genesis description of the pre-Flood violence that filled the earth include dinosaur violence as well as human violence? Surely "all flesh" included these prominent animals. The promised judgment inundated the entire earth, the domain of all its inhabitants.

The lateral extent of the strata in which Sue and many other large fossils are found indicates that the depositional events were often explosive volcanic conflagrations of semi-continental extent, all in a water environment. Sue and many other large fossils were buried in flowing volcanic mud under raging waters that covered a large area.

The puncture wounds in "Sue's" jaw were left over from a violent encounter with another *T. rex*. They healed, but she must have suffered from excruciating pain from that point on. Note that the femur, or thigh bone, of Sue had been broken. It naturally reset, but could bear little weight. Sue recovered from this violent attack, but for the rest of her long life suffered horribly. Similarly, several of her ribs had been broken. They healed naturally, but incompletely. Every breath brought her pain. (Location: South Dakota. Dated as Cretaceous.)

And now dinosaur fossils are found on every continent, in catastrophically deposited widespread strata, along with many other fossil types that were also rapidly deposited victims of violence.

Mammals

It has long been taught in evolutionary literature that during the Mesozoic (the dinosaur era), mammals were rather small and insignificant in comparison to the dinosaurs. Recently, however, fossils of two large mammals discovered in China have challenged this oft-repeated claim.

The fossils are thought to date from about 130 million years ago, long before evolutionists believed large mammals began to roam. Named *Repenomamus robustus* and *Repenomamus giganticus*, the former was a meat eater and had the remains of a dinosaur in its belly. So much for fanciful stories about the long ago, unobserved past, when "dinosaurs ruled the earth."[8] On some occasions, at least, dinosaurs were obviously prey for mammals.

Evolutionists are uncertain regarding many aspects of mammalian evolution. "The presence of hair in mammals may seem a trivial matter, but the exact timing of the origin of hair in mammals is unknown because the fossil record of the evolution of hair in mammals is exceedingly sparse."[9] The evolutionary development of major traits is often so far beyond current understanding that the only recourse is a "story."

Decades of imaginative "just so" stories, now enhanced with vivid computer animation, have shaped people's thinking and, to a great extent, scientific thought. Yet evolutionists demand that their changing "stories" be accepted at each stage. This is not meant to deny evolutionists the opportunity to discover new information, and the freedom to change their opinions if warranted, but it is a lesson for the rest of us. Sometimes scientists are completely wrong about important matters, their wild speculations and storytelling found wanting. Students should never be intimidated into believing things they suspect might be wrong. And scientists should display more humility in matters in which their information is incomplete, and for which they have only a tiny fraction of the relevant data. Scientists are limited in their knowledge and certainly not infallible in their evaluation.

Mammal and other vertebrate fossils compare poorly in number to the multitudes of invertebrate fossils found. As vertebrates ourselves, we are naturally more interested in other vertebrates, and they have been widely used to indoctrinate students in the evolutionary way of thinking. Space does not permit a thorough critique of all mammals, many species of which probably date from the catastrophic centuries following the Flood. Our attention turns to three fossils successfully used by

evolutionists—horses, whales, and man.

The Famous Horse Series

Horse evolution prominently appears in museums and textbooks as the supreme example of the evolution of one body style into another. All students have seen the "horse series" sketches, which show how small browsers named *Hyracotherium*, with four toes on the front and three on the rear (referred to as 4/3), changed into the large one-toed horse of today (1/1). Intermediate steps included *Mesohippus* with three toes (3/3), only one of which touched the ground; one-toed grazers named *Merychippus*; the larger *Pliohippus*; and finally our modern horse, *Equus*. Along the way, it acquired high-crowned molars and other adaptations.

Often the concept of a "missing link" in the layman's perception is one type of animal that displays traits that are halfway between two others, linking them together in an ancestral/descendant relationship. But there is more to it than that. As it is commonly represented, the horse series more closely resembles the proper scholarly concept of an evolutionary series: a sequence of small changes in body structure (morphological series) found from low to high in the geologic column (stratigraphic series). Morphologically, the horse series shows several sequences of small changes from 1) 4/3 toe number, changing to 3/3 with toe 2 raised, and further changing to 1/1 in the modern horse; 2) small body size to medium size to large size today; 3) short face with no tooth gap (diastema) to long face as in today's horses, with gap between front and rear teeth; 4) teeth for eating leaves and bark (browsers) to teeth for eating "gritty" grasses (grazers).

These gradual structural changes (morphological series) were originally thought to be spread out from lower to higher through the Cenozoic (Tertiary and Quaternary systems), a stratigraphic series covering "65 million years" of evolutionary time. That may seem like a huge amount of time to accomplish relatively minor changes in body style (compared to "only" three million years to make the more dramatic body changes from ape to man), but at least the horse series shows what an evolutionary series should look like (versus a single "missing link"). Unfortunately for Darwin's followers, further research once again exposed serious flaws in the horse story.

The little 4/3 browser that started it all is often incorrectly called *Eohippus*, meaning "Dawn Horse." Actually, the fossil had been discovered earlier and named *Hyracotherium*, meaning "hyrax mammal," because it looked so much like the modern hyrax, an animal called a rock badger or coney in the Bible. Its descendants still live today; there is no evolution here, just stasis.

Scientists who study fossils are trained to notice differences and similarities between them, allowing for their proper classification. However, the agenda of many evolutionists leads them to focus on the changes neces-

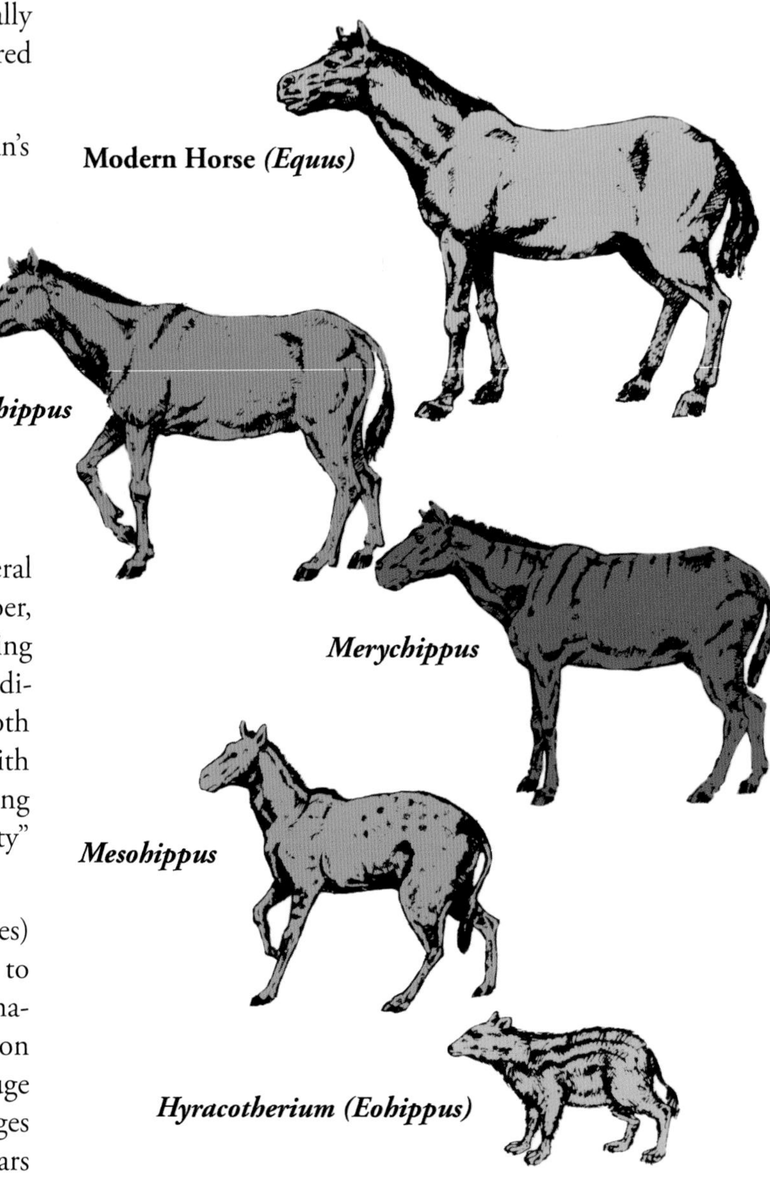

Evolutionary theory claims *Hyracotherium* evolved into horses, elephants, manatees, and other mammals. The transition as shown looks impressive and was thought to support straight-line evolution, but when it was recognized that the different types sometimes appear in the same strata, this view disappeared. There is variety among horses types, but not evolution between the different types.

sary to bridge the differences, and to imagine stories about how an organism acquired the needed genetic mutations that could then be preserved by natural selection. They often employ false action verbs and computer graphics to describe how certain animals, represented only by their fossils (dead things), "chose" to change in a particular way over time to meet a particular need. For example, grazing teeth were needed, so the horse somehow acquired them. It is almost like natural selection has a mind and decided to change in a certain fashion.

Early treatments considered that evolution progressed in a direct line from one type to another, and museums displayed horse fossils in that sequence. But in recent decades, evolutionists have disavowed this straight-line view of directed evolution in general, and of horses in particular. They now do not consider any particular form to have been the goal of "non-directed" mutation and natural selection. Once freed to examine the data without this "directed" overprint, scientists recognized that differences among horses had been concurrent and types had often overlapped each other.

Of course, a variety of modern horses exist today, with adaptations that enable them to cope with widely varied environments. All horse species hybridize in nature. Strains of the domestic horse may be bred to accentuate a single trait, such as tiny horses that are about as large as a dog. Actually, the many breeds of horse produced by human intervention show wider variation than the natural species, but variation is not evolution. If man

A rare, fully articulated fossil of *Hyracotherium*, found in the Green River Shale. About the size of a small dog, the claim is made that this creature evolved into several large mammals, including horses, elephants, rhinos, tapirs, and manatees. Its fossils are found in Tertiary sediments. In pre-Flood days, many mammals underwent extensive variation within the limits of their kind or basic type, but not evolution into different kinds. Variety is not evolution. Usually, when evolutionists give an example of "evolution," they only cite evidence of minor variation within limits. (Location: Green River, Wyoming. Dated as Eocene.)

had selectively bred horses for face length like they did with dogs (bulldogs or greyhounds, for example), the variation would exceed that among fossil horses.

During the same time period that evolution claims some of the descendants of the little *Hyracotherium* supposedly developed into full-blown horses, some of its other descendants were thought to become the manatees, elephants, and several other large mammals. Yet some individuals in the genera (the hyrax) persisted relatively unchanged throughout the same supposed great span of time.

Fossil horses of all the various so-called evolutionary "stages" are found in the same strata intervals. In life, they were contemporaries, perhaps living in a patchwork of adjacent or overlapping environmental zones. They could not have an ancestor/descendant relationship. This is especially true for the "horse series." The original fieldwork that established this fossil folklore was carried out in the John Day country in Oregon, but there fossils of the three-toed grazer *Neohipparion* (very much like *Merychippus*) have now been found with *Pliohippus*. In the Great Basin area, *Pliohippus* has been found with the three-toed *Hipparion* throughout the time supposedly represented. Evolutionists freely admit this situation, and to their credit often attempt to correct the misconceptions, but still the horse series appears in textbooks and museums. It is now thought that horse evolution as recorded in the fossils follows a meandering and overlapping pattern and not the classic "tree" at all. Indeed, the different horse fossils we find may reflect environments, not changing species. A one-toed grazer cannot precede its three-toed, browsing ancestor.

Remember that any three fossils can be placed in a line and an evolutionary story told about the transformation of one into the others. And a different story could be told if the fossils were arranged in a different order. With or without its favorite stories, evolution remains the dominant origins myth for Western culture, modifying itself to fit all observations. If the theory can accommodate any possibility, it is a weak concept indeed, full of sound and fury that signify nothing.

Whales

The story that whales evolved from a land-dwelling creature looms large in evolutionary lore. Whales are, of course, marine creatures that are usually of immense size, but they are mammals rather than fish. Thus, according to evolution they must be directly related to terrestrial mammals, which evolved from fish. Did they evolve from a terrestrial ancestor? If so, which one? Colbert noted their sudden appearance:

> Like the bats, the whales (using the term in a general sense) appear suddenly in early Tertiary times, fully adapted by profound modifications of the basic mammalian structure for a highly specialized mode of life. Indeed, the whales are even more isolated with relation to other mammals than the bats; they stand quite alone.[10]

The world's "oldest" bats are found in Fossil Butte National Park. Claimed by evolutionists to be around 50 million years old, they are scarcely different from bats living today. There are no fossils found of animals that are gaining bat characteristics. An animal was either a bat or not a bat. What is seen in the fossil record is stasis.

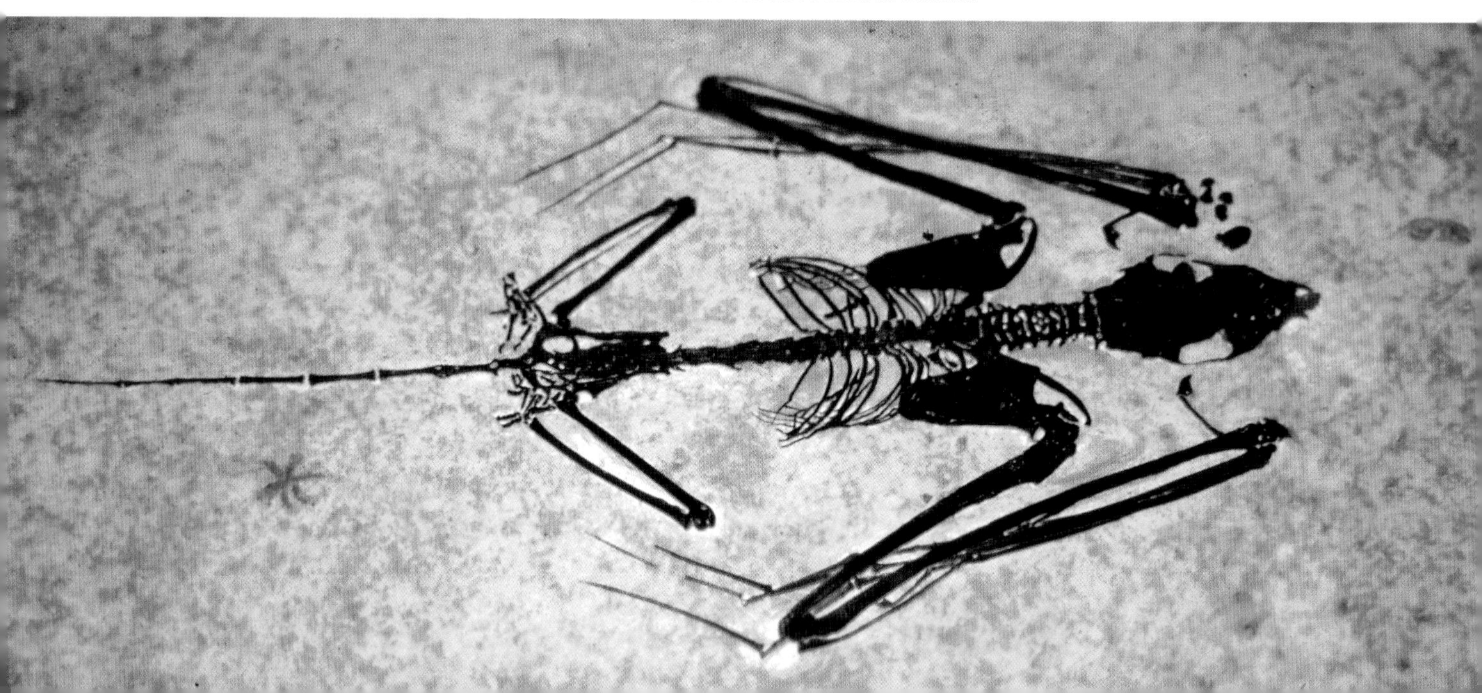

To the uninitiated, the entire claim of whale evolution from a land creature would seem ludicrous on its face. What natural pressures would force a highly adapted land mammal to undergo impossibly complex and radical changes (e.g., in reproduction, respiration, locomotion) for an aquatic existence? These marvelous beasts, so very different from any other living creature, deserve recognition for their unique traits, and should not be minimized for any faint similarities with land creatures. Conservation efforts rightly attempt to "save the whales" for the sense of awe they instill in us, as well as for their importance to the world's ecosystem. They stand on their inherent value, not just on their relationship to other mammals and man.

Darwin hypothesized that bears had been the progenitors of whales, arguing that bears still venture open-mouthed into the water in search of fish.

> I can see no difficulty in a race of bears being rendered, by natural selection, more and more aquatic in their structure and habits, with larger and larger mouths, till a creature was produced as monstrous as a whale.[11]

The bear–to-whale idea has long since fallen out of favor. Throughout much of the twentieth century, the even-toed ungulates (the herbivorous, hoofed artiodactyls, including cattle) were championed as the whale ancestor. During the last few decades of the 1900s, wolf-like carnivores—the mesonychids, which supposedly lived 60 million years ago—gained in popularity. In 1988, a book on aquatic mammals was cautious, saying:

> The origins of cetaceans are still uncertain. Many aspects of cetacean anatomy...indicate little about ancestry, but biochemical and genetic studies suggest cetaceans are related to hoofed animals...the Mesonychidae. Cetacea probably arose from a small form...we can speculate that one of these possibly fish-eating animals became adapted to feed on abundant food in shallow waters.[12]

Benton stated only that "mesonychids are probably close relatives of whales,"[13] while Colbert said, "The mesonychids are now placed in their own order close to the whales."[14] Other evolutionists maintain that a hyena (*Pachyaena*), or perhaps *Sinonyx*, a cat-like creature, gave rise to modern whales.

Once DNA comparisons were fine-tuned, Jessica Theodor of the University of Calgary and Jonathon Geisler of Georgia Southern University suggested that the hippopotami were the ancestors of whales. Hippos evolving into whales? These creatures are designed to eat only vegetation, as seen by their grinding type of teeth. But whales are meat eaters—even the krill-eating baleen whales. A similarly daunting problem is the fact that according to evolutionary dating, whales have been around five times longer than hippos. How could the descendant predate its ancestor? Add to this the opinion of J. G. M. Thewissen and his group, who insist that hippos are more closely related to pigs.

In December 2007, the fossil of an herbivorous deer-like creature about the size of a raccoon was found. This now commands a sizable following as the ancestor of carnivorous whales. *Indohyus* was an artiodactyl (a group that includes pigs, hippos, and giraffes) and supposedly lived in southern Asia "48 million years ago." The breaking Associated Press story began, "It sounds like a stretch." Non-Darwinists certainly agree! The *Indohyus* report by Jason Palmer in *New Scientist* states:

> The research also challenges the idea that cetaceans—the order that includes whales, dolphins, and porpoises—split from their land dwelling forebears and returned to the water to hunt aquatic prey....This suggests that *Indohyus* was a shallow water wader already, and had not returned to the water simply to hunt live prey.[15]

Obviously, this is a direct contradiction to the traditional evolutionary notion of terrestrial animals moving to the ocean. At each stage along the way, advocates' "just so" stories claim to have the proof, only to see their ideas replaced by newer ones. Dead ends and false starts are the rule when addressing the elusive subject of the whale's alleged terrestrial ancestor. Obviously, the evidence is ambiguous at best. To recap, paleontologists have suggested that various creatures—including mesonychids, the hyena (*Pachyaena*), *Sinonyx*, and now *Indohyus*—were in the whale evolutionary line. Perhaps evolutionists ought to question the entire idea of whales evolving from a land creature of any sort.

In considering evolutionary claims, however, it is good to go back to the important questions recommended for every such assertion: "What do you mean by that?" and "On what evidence do you base this conclusion?" Whales provide a good example of how to use and profit by these questions.

The first question is really a category of question and a method of questioning. It can be formulated in many ways, and should perhaps be politely asked as a series of questions. The query "What do you mean by that?" or "Can you help me more fully understand that statement?" or "Does this imply that…" can lead to the heart of the matter. It simply is a request for clarification and more information. Consider the following list of possible questions, which could be much extended. Each will be amplified in turn. Remember, they must not be asked without fully understanding the question, nor should a student attempt to ask them all or ask them at an inappropriate time. They are a type of question, not a list of "stumpers" for the teacher.

What do you mean by that?

1. Do you mean the legs shortened and mutated into flippers?
2. Did the small tail become the fluke, the whale's broad, powerful swimming tail?
3. Do you mean that the nostrils migrated to the top of the skull to become a blowhole?
4. Do you mean the brain case thickened to withstand high water pressure when diving?
5. Do you mean the hide lost its hair and developed a thick layer of blubber for warmth?
6. Did mother whales, which are mammals, acquire the ability to give birth and nurse underwater?
7. Did the lungs and other organs adapt to the new environment, or were they replaced by a different design?
8. Do you mean that the teeth of some were replaced by baleen? Where did baleen come from?
9. Do you mean that each of these changes happened gradually? In other words, did the legs shorten gradually?
10. Did the changes occur as mega-mutations?
11. Did all of these adaptations occur simultaneously?
12. Who is wrong—those who claim that the whale came from a wolf-like creature, or those who believe it was from a hippopotamus? What about those who think it was a deer-like animal?

Let's look at possible answers to the questions.

1. Do you mean the legs shortened and mutated into flippers?

The standard answer is yes. But is this process as simple as seems to be suggested? Whether it was a fish-eating bear, a vegetarian grazer, or a carnivorous wolf, any ancestor would have had to undergo a radical alteration for life in the sea. To begin with, random mutations in its genes would have to result in shorter, mutated legs, which over the generations underwent similar mutations that resulted in still shorter legs. This would not happen just by habitually living in the sea, but by random mutations. According to this scenario, the legs eventually disappeared, and in their place were gradually lengthening flippers, produced by other mutations. Today, whales swim primarily by thrusting their tails up and down and by arching their spines, not by using their flippers for transportation at all. Where did they gain this ability?

2. Did the skinny tail of the ungulate (cattle-like creature) or wolf or hippo become the fluke, the whale's broad, powerful swimming tail?

How? Explain this to me. The fluke on a whale is such a broad, powerful, and necessary organ. What feature on any land animal could mutate into this? All land animals' tails operate from side to side, but the whale's movement is up and down. This requires different muscles and skeletal connections in the vertebrae. All must be in place and operational for the fluke to work. These changes would require separate, coordinated mutations. How could the whole suite of changes occur together?

3. Do you mean that the nostrils migrated to the top of the skull to become a blowhole?

They must have done so. Some whales have two blowholes, more like the nostrils on a terrestrial animal. In some, the nostrils are in an intermediate location, but always fully functional for the whale's behavior. Changing from one breathing mechanism to the other would take many, many, generations, but there is no evidence of this. Interestingly, the whale has a mechanism to close its nostrils to keep out water, but few land animals have this ability. To add this function requires the necessary flap, the muscles to control it, and the nervous system to activate it. Holes in the top of the head also imply cranial alterations. Since breathing is necessary for each generation, how quickly did this change occur?

A fossilized bone from the ear of a whale. Whales can hear extraordinarily well underwater. The bones it utilizes are similar to those of land mammals. Evolutionists propose this as a major point supporting the idea that whales evolved from land mammals. But the corresponding bone in any land mammal is very tiny, not at all like the robust bone in whales. This hardly seems good evidence for the general claim.

4. Do you mean the brain case thickened to withstand high pressure when diving?

Some species of whale dive to extreme depths in search of food. There, the water pressure is unthinkably high and would crush the skull of any ancestral land animal. This would require many successive mutations to the ancestral type, each yielding a thicker skull than the previous generation. According to natural selection, each mutant had to be somehow favored in order to be selected. It then had to survive to reproduce more mutants, while those whales without the favorable mutation went extinct. These mutations are distinct from those required for the other changes already mentioned, so how could there be enough generations for them to accumulate? How long is a whale generation? They are not like fruit flies, which reproduce every couple of weeks. How many generations are possible in the relatively few million years evolutionists postulate this process took place, assuming each generation got just the right mutation and in just the right order?

5. Do you mean the hide lost its hair and developed a thick layer of blubber for warmth?

Terrestrial mammals usually have hairy hides, to one degree or another. A whale is smooth and hairless. The loss of all the hair on a mammal is sometimes quite harmful. How could such a mutation be an evolutionary step forward? The blubber on a whale is a very complex and a necessary source of nutrition and warmth. It is not comparable to the fat layer in normal mammals. It seems to be something different from anything that had gone before. Where did it come from? Did something on a land mammal change to become the necessary layer of blubber? If so, what?

6. How did mother whales acquire the ability to give birth and nurse underwater?

Birth among air-breathing whales requires immediate access to air for the offspring. It cannot work like a land animal, where the young simply drop to the ground and breathe. A new system, complete with adapted body parts and new habits, must be in place from the earliest days the animal took to the water, or there would be no next generation. A female mammal's nipples are normally on her underbelly. When a whale calf nurses, a special skin flap covers the calf's mouth to keep it from drowning, while special muscles in the mother rapidly force milk into it. What feature on the whale's ancestor was modified by mutation to allow this?

7. Did the lungs and other organs adapt to the new environment?

Nearly all the major organs in whales function somewhat differently from those in their terrestrial counterparts. For instance, land animals drink fresh water, but whales need to be able to filter or handle the excessive salt in seawater. Furthermore, their lungs needed to increase their volume and oxygen exchange many times over to allow for long, deep dives. Their eyes and ears had to toughen to withstand increased underwater pressure, and their ears had to be able to pick up both airborne and underwater sound waves. How did the intermediate forms survive until all the necessary parts were in place?

8. Do you mean that the teeth of some were replaced by baleen? Where did baleen come from?

Did baleen whales and toothed whales evolve simultaneously from different sources, or did they split off

be correct, but support for it must be unassailable. There must be a reasonable lineage documenting such a transition.

Note that both types of creature possess hard bones that can be fossilized under the right conditions, and fossilized remains of both are found in sufficient numbers. Are there intermediate fossils? If so, do they fit the expected transitions between these very different animals? Is the ancestor/descendant relationship of sufficient quality to make the "outrageous" claim stick? Given the time it takes for large animals to mature, were there enough generations for natural selection to act and to also eliminate all the individuals that had not received the proper mutations?

A recent treatment admitted: Maybe not. "The origin of Cetacea [dolphins, porpoises, and whales] has been an enduring evolutionary mystery since Aristotle."[16]

Evolutionists once lamented the lack of fossil evidence for whale evolution, but now they brag that the evidence has been discovered and is convincing. Most up-to-date museums consider fossils documenting whale evolution to be one of their main exhibits. Just how good is the evidence?

Unfortunately, when searching for the current view among evolutionists, you will find that there is no consensus of opinion. Often, the statement is made that the case for whale evolution is strong, but it is obviously not strong enough to yield agreement among evolutionary "experts." Every new find replaces the prior certainty with a fresh one. It is quite instructive to look at the recent discoveries in order to see just how much credence is due these claims.

The classic 1966 college textbook on vertebrate paleontology by Alfred Romer acknowledged the scanty and speculative nature of the evidence for whale evolution, and then stated:

> Many features of their structure suggest their origin as a branch of primitive carnivorous or pre-carnivorous stock which had taken up a fish-eating life, but a number of important modifications had already occurred in the primitive Middle Eocene Protocetus and Prozeuglodon of the Upper Eocene....Skeletal remains are rare."[17]

Romer had nothing to say about the identity of any possible whale ancestor.

Updating Romer's work, a colleague wrote a most enlightening textbook published in 1985, and refreshingly admitted evolution's weaknesses:

> As with most tetrapods secondarily modified for aquatic living, ascertaining the terrestrial stock from which the whales came is exceedingly difficult...paleontologists have had to rely upon indirect reasoning to identify the mammalian group most probably ancestral to the first whales. Observing that the archaeocetes were fully adapted for swimming by the Eocene epoch, investigators recognized that similarities between whales and animals which appeared then and later, like the modern ungulates, sirenians, and pinnipeds, would have little or no phylogenetic significance. R. Kellogg pointed out, and others agreed, that the transformation of a terrestrial mammal into a completely oceanic one was so great that the process must have begun at least as long ago as the early Paleocene and possibly even before that time, at the end of the Cretaceous period.[18]

Stahl pointed out that the large, carnivorous ancient whales, the archaeocetes, could hardly have evolved from a rather tiny insectivore source. She referenced Van Valen's work proposing the mesonychids—primitive wolf-like carnivores that had teeth somewhat similar to the much larger teeth in early whales—as a more likely ancestor. This train of thought continued for several years. None had transitional fossils to guide them. All treatments of whale evolution speculated on the role played by *Basilosaurus*, an 80-foot-long, snake-like creature. While acknowledging that it was probably an early whale, Stahl insisted that "the serpentine form of the body and the peculiar serrated cheek teeth made it plain that these archaeocetes could not possibly have been ancestral to any of the modern whales."[19]

The fifth edition of Colbert's standard paleontology textbook repeated this stance in 2001.

> These mammals probably arose in early Cenozoic times from primitive carnivore-like mammals known as mesonychids, which are now usually classed as ungulates....The whales (using this term in a general and inclusive sense) appear suddenly in early Tertiary times, fully adapted by profound modifications of the basic mammalian structure for a highly specialized mode of life....Therefore, it seems evident that the whales, having separated from their meso-

nychid ancestors at an early date, enjoyed at the outset a series of extraordinarily rapid changes that made them by middle Eocene times well adapted for life in the ocean.[20]

After renaming the Mesonychidae as a new order, the Acreidi—which contained only the one family, the Mesonychidae—he says, "Evidently, the Acreodi occupy a position close to the ultimate ancestry of the whales."[21]

All of the textbooks agree that the mesonychids lived on land and looked rather like a wolf or hyena. It was not whale-like at all, except that some of its fingertip-size teeth reminded evolutionists of much larger whale teeth, and features in the whale's inner ear were reminiscent of that in the mesonychids. This tenuous association supposedly invalidated earlier suppositions that a bear or browsing ungulate had evolved into a whale. But was this candidate sufficient? Let's continue.

In 1979, a fossil found in Pakistan gave evolutionists something to rally around. Phillip Gingerich, the discoverer, made worldwide headlines, which heralded the first "indisputable" evidence of whale evolution.[22] Dubbed *Pakicetus,* it had teeth "resembling not only mesonychid teeth, but also the teeth of middle Eocene cetaceans."[23] But a few teeth is about all it had. This "diagnostic" fossil consisted only of a partial skull and jaw and the few teeth, but no bones from the rest of the body. It was found in river sediments, yet it was considered "fully aquatic," the "most primitive cetacean [whale] known."

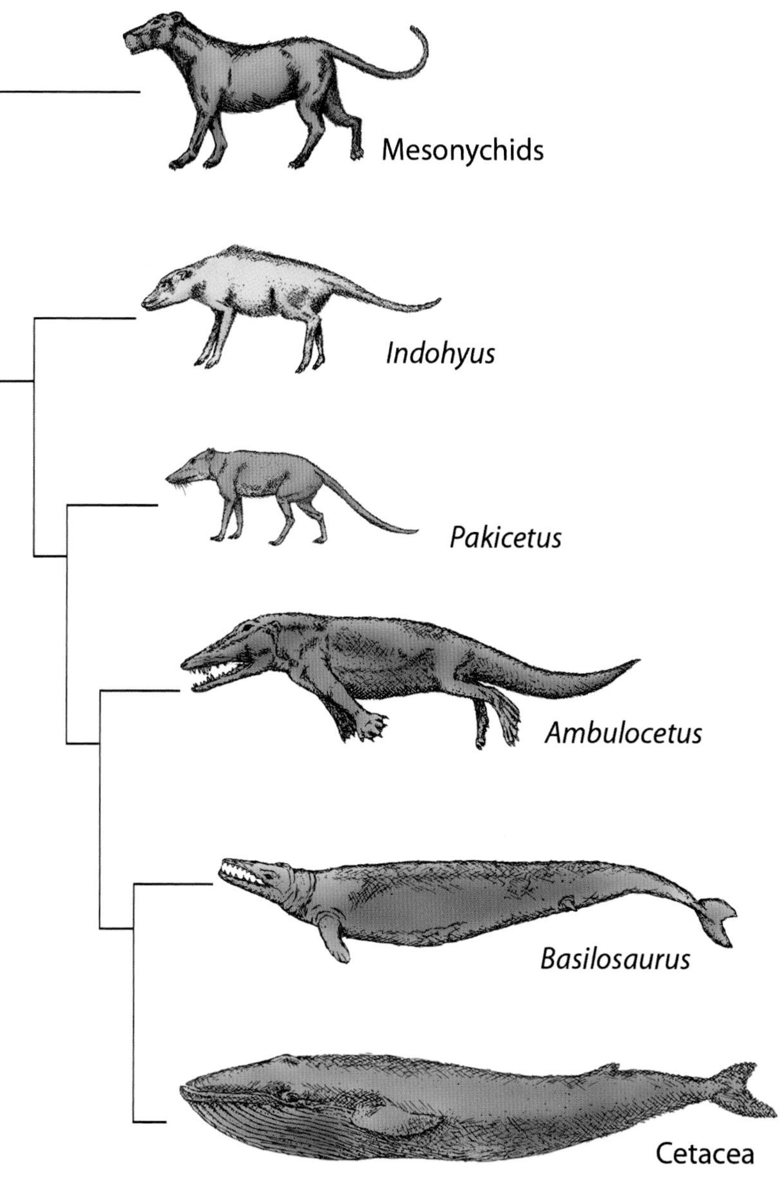

The proposed evolution of whales from a land animal supposedly involved major changes over several million years. This transition should be well-documented in the fossil record, with many in-between forms. A few fragmentary fossils have been found and they are interpreted in the most favorable light possible, but all are partial and must be reconstructed. Where good fossil evidence is found, a case for evolution cannot be made.

Illustrations depicted a swimming creature pursuing fish for food. It had webbed feet and a streamlined body, a portrayal based on a leap of faith that was totally unsupported by any fossil evidence other than a few teeth. Trouble with the reconstruction started when Professor Hans Thewissen and others discovered additional bones of *Pakicetus.* Once they had better information, its reconstruction looked nothing like the prior illustration.[24] The creature was typically a land-dwelling mammal, with essentially the appearance of a wolf-like carnivore, and was not much different from the mesonychids. The much more extensive skeletal remains showed that this creature was a true land mammal, able to run in the normal fashion. While it may have waded in the shallows, it was not fit to be a habitual swimmer. Artists' reconstructions had given the wrong impression, but the bones proved it was not the transitional form that had been hoped for. Unfortunately, the improper illustration still appears in many treatments of whale evolution, and advocates continue to call this wolf-like carnivore—which could run and live on land only—a cetacean.

Thewissen, although retaining his belief in whale evolution, assesses it thusly:

> Pakicetids were the first cetaceans, and they are more primitive than other whales in most respects. In fact, they did not look like whales at all, and did not live in the sea. Instead they lived on land, and may have fed while wading in shallow streams.[25]

How could he label such a creature "a whale"? Creationists would agree with his evaluation of *Pakicetus* in general, but they would vigorously object to calling a wolf-like land-dweller "the first cetacean" in an effort to salvage the evolution story. No matter what its tiny teeth looked like, no matter how much they remind evolutionists of a large whale's teeth, *Pakicetus* was not a whale, and students should not be deceived or intimidated into considering it so. Evolution should have better evidence, and not have to use this self-serving claim to mask such an outrageous hypothesis. Actually, the *Pakicetus* teeth, no bigger than the fingernail on your little finger, do not look much like a whale's tooth, so it is not really the evidence needed. Simply noting similarities between mammals, of which there are many, does not prove relatedness.

All the possible fossil "whales," the so-called archaeocetes, discussed up to this point do not seem to have been whales at all. They were all fully land-dwellers with legs. Similar fossil creatures are named the wolf-like *Elimeryx*, the otter-like *Rodocetus*, and others. The same discussion would apply for them as to the adequacy of the fossils. Despite the fact that some of their characteristics are shared by many mammals, they were simply not whales in the normal use of the word.

Demonstrating the fragile status of evolutionary transitional fossils, *Indohyus* burst onto the scene in 2007. This too was completely a land mammal. Not at all whale-like, it was the size of a fox or raccoon. It looked something like a miniature deer, yet it dethroned *Pakicetus* et al from consideration, at least for the present. Its bones were somewhat thicker than other mammals, suggesting to Thewissen that it spent much of its time wading in water. Study of the chemical composition of its teeth indicated it was a plant eater, disproving the speculation that land animals ventured into the water for fish or clams. According to Thewissen, "*Indohyus* is a plant eater, and already is aquatic."[26] One might wonder if changing the speculation for the whale's ancestor from a cow-size grazer to a wolf-size carnivore to a hippopotamus to a tiny deer-like creature might not be going in the wrong direction.

The next candidate that supposedly spanned the transition from land-dwellers to true whales was *Ambulocetus*. It is discussed in every treatment, but was it sufficient to fill the gap? Remember, fossils can always be displayed in any desired order, in any linear array, and an evolutionary story can be told about the transition between adjacent members. But did one really change into the other, and is the evidence given sufficient to

Crocodile-like phytosaurs were giant reptiles. They looked like dinosaurs, but they lacked the right hip structure. Their fossils are often found in the same strata as dinosaurs and show persistent stasis, not evolutionary change. (Location: West Texas. Dated as Triassic.)

make the story stick? Does this particular proposed whale series actually reflect truth?

The formal name *Ambulocetus natans* means "walking whale that swims," a name that promotes its interpretation as an evolutionary transition. It does seem the animal had an amphibious lifestyle, for it evidently spent time both on the land and in the sea. Thewissen discovered the partial and scattered specimen in 1994. It was immediately accepted as the long-sought missing link, and several beautiful interpretive paintings were produced of the creature in action. Several years later, more of the skeleton was recovered and its paintings received a significant makeover. Unfortunately, they still presented more than could be known about what the animal looked like and how it lived.

Most of the skull and vertebrae were recovered, and some of the ribs. Also present were some of the foot bones, and these indicated that it was a strong swimmer that propelled itself with powerful kicks. The pelvic girdle and most of the legs were not recovered, and much of what was there was shattered. The pelvis and legs are the bones that hold the crucial information for determining the creature's lifestyle, and especially how it moved. Through this, it was determined that *Ambulocetus* occupied an ecological niche like crocodiles inhabit today, wallowing in the shallows and preying on luckless land creatures, and probably shallow-water fish and turtles. The questions remain: Was it related to whales, and how was it related to its supposed terrestrial ancestors? Maybe it was just related to crocodiles.

It was also claimed that the ear was reminiscent of land animals, but wouldn't it be? Other animals not thought to be related to whales have similar features. Besides, an ear does not make a whale, and a deep-diving whale needs a much more distinctive ear.

The most important issue at hand concerns the legs and their claimed loss in whales. Some whales have claspers where legs would be. These are useful in copulation, but are much smaller than actual legs and do not protrude beyond the skin. They are obviously not "legs" and are quite necessary for reproduction. *Ambulocetus*, on the other hand, had true legs, not tiny legs or vestigial legs. Its lifestyle was probably similar to that of modern sea lions—quite adept in the sea, but able to get around on land. Sea lions are certainly clumsy when they come ashore for sleep or to birth their young, but they are not half land-creature and half marine-creature. Only an attempt to artificially bridge the gap between land and marine denizens could lead to the calling of this sea lion-like animal a "whale that walked." Sea lions are not whales! They are what they are and are quite successful at it.

Remember, an animal that lives in two environments at different times of its life or to carry out varied functions of its life is not necessarily transitional. Each such animal is marvelously designed to move back and forth across the terrestrial/aquatic or other boundary. This shows precise forethought, not adaptation and mutation.

Fossils somewhat similar to *Ambulocetus* have also been touted. Named *Gandakasia* and *Himalayacetus,* the discussion above applies to them as well. None of them are whales or give sufficient evidence that they are related to whales. Only in the context of the assumption of evolution do these fossils take on feigned significance. In science, "seeing is believing," but in evolution, "believing is seeing." It takes a lot of believing to see an evolutionary thread through the scattered, shattered fossil fragments that serve as the basis for so many different "just so" stories and illustrative paintings.

Next in the supposed whale family tree is a genuine sea creature named *Basilosaurus.* This fossil was discovered in the early 1800s. Due to its long narrow body (up to about 80 feet) and multiple vertebrae, it was originally considered a reptilian sea snake. The recent interest in whale evolution has resurrected this creature, and it has been elevated it to iconic status by many experts—but not all. Note the testimony of Dr. Barbara Stahl in her 1985 paleontology textbook:

> The later archaeocetes...the monstrous 60-foot *Basilosaurus* had lost the sacroiliac articulation, lengthened the lumbar vertebrae, and eliminated the bracing of one neural arch upon the next so that the long trunk and tail region attained maximum flexibility. The serpentine form of the body and the peculiar serrated cheek teeth made it plain that these archaeocetes could not possibly have been ancestral to any of the modern whales.[27]

Several nearly-complete skeletons of *Basilosaurus* have been found, including one with a tiny leg nearby. The leg was only about six inches long, compared with the 60-foot-long creature. It was not attached to the vertebral column, making the use of this "leg" (if indeed it was really part of the animal) uncertain. Modern snakes have similar tiny features used as reproductive claspers, which likewise do not extend beyond the skin. It is often drawn in evolutionary treatments with a much thicker (more whale-like) body and with a much longer (less serpent-like) leg than is warranted from the fossils. The question remains: Was this a whale or even a whale ancestor?

Many of today's evolutionists answer with a resounding "No!" They feel that *Basilosaurus* may have been an early whale relative, but its differences set it well apart from modern whales. To scholars, it was an extinct branch of whales' evolutionary tree, but not in the line of later whales. Unfortunately, *Basilosaurus* is still depicted in most of the textbooks, museums, and TV specials as an important link in the chain. Some recently have placed more importance on the similar sea creature called *Dorudon,* a porpoise-size animal approximately one-tenth the size of *Basilosaurus.* It is sometimes deceptively drawn as nearly the same length and proportions as *Basilosaurus*, but the differences are striking.

We may never know the whales' "family tree" based solely on the fossil evidence. *Basilosaurus* and *Dorudon* may have actually been mutant whales, having undergone extensive modification in the years immediately following the Flood, adapting in response to the new and different environments in which they found themselves.

Numerous amphibious creatures are exquisitely designed to move daily from land to water and back, but here is the take-away truth: Whales are whales, and land creatures are land creatures. As you can see, there is insufficient evidence to establish the outrageous hypothesis that whales evolved from land creatures. Imaginative action stories, art, and computer animation must be employed to "sell" evolution to the public (or science classroom)—the evidence could never do it.

Human Ancestors

Horses, whales, and dinosaurs—all are prominently used by evolutionists to promote their otherwise poorly supported theory. But the ultimate poster boy for evolution in many people's minds is the real "missing link," the one between man and the apes. An overzealous search for links has produced some of the most embarrassing events in the history of science. It illustrates the intense desire to rank man in the animal kingdom, thus proving that evolution is more of an idea about "who we are" than "where we came from."

Perhaps the most famous paleontological hoax in science was the so-called "Piltdown Man." A devious fraud perpetuated by leading evolutionary scientists, it was a deliberate attempt to furnish the data that honest paleontology could not provide. The skull cap and jaw found (perhaps planted by one of the scientists involved) near the English village of Piltdown in 1912 were hailed as the perfect link "halfway" between ape and man. It was optimistically named *Eoanthropus dawsoni* (Dawn Man of Dawson, its discoverer). The fake fooled virtually all the world's leading evolutionists for over forty years. It was used for decades to indoctrinate students with the idea that this fantastic find falsified the biblical view of mankind's special creation.

In the 1950s, the mismatch of minerals in one bone and the surrounding rock caused scientists to look more closely at those hallowed "fossils." They turned out to be a bit of human skull with key parts missing and an ape jaw with the teeth filed to look more human. Both were stained to make them look older and then buried with other fossils, including one imported from Africa. The evolutionary worldview had so strongly influenced the scientific research that for forty years observers had seen what was not there and failed to see what was there.

While Piltdown Man was being touted as a "fossil link" from ape to man in the early 1900s, Darwin's second book (*The Descent of Man*, 1871) and the work of "Germany's Darwin" Haeckel suggested to some evolutionists that Australian aborigines and African Negroes were subhuman "living links." Haeckel, who was caught mis-drawing human embryos to make them look more animal-like, drew aboriginals as animals in trees and claimed that they could not count as high as dogs or horses. Hitler based his beliefs on the racial inferiority of Jews and the superiority of Aryans on his interpretation of Darwinian ideas.

These "false facts" of early evolutionary enthusiasm were misused to prove evolution to "ignorant" Christians at the famous Scopes "Monkey trial" in 1925 (an event radically distorted in remakes and reruns of the famous film *Inherit the Wind*). Modern evolutionists repudiate such claims, of course, and several other ape-man candidates have come and gone since then. Today, the two most popular fossils claimed to link apes and man are Neanderthal and "Lucy," and these merit our special attention.

Neanderthal

In the mid-1800s, strange bones that closely resembled human bones were found in the Neander Valley in Europe. At the time, they were hardly more than an interesting curiosity. In the early 1900s, however, these fossils came in handy as many were seeking evidence to support the theory of human evolution from the animals. The skeletons had a stocky build and certain unusual characteristics, such as a sloping forehead, while their habitations indicated a decidedly primitive lifestyle. What better example of a half-man, half-ape could be expected? These interesting creatures have been the focus of intense study ever since, and there is no consensus yet among scholars as to their place in human history. Are they part of the modern human variety, a side branch on the evolutionary tree that went extinct, or were they assimilated into modern human groups?

DNA retrieved from some of their bones shows Neanderthals to be on the genetic fringes of the modern human range, but still within the human range. They are closer to the norm of all ethnic groups than are modern Australian aborigines, which, although distinct in several ways, are part of an undoubtedly human people group. As it turns out, the average brain size of Nean-

This imaginative illustration of "monkey to ape to man" often appears in treatments promoting human evolution. But each basic type exists today, and they do not hybridize in the present. How do we know one evolved into another in the unobserved past? Many frauds have been perpetuated and mistakes made in trying to establish this series showing an animal heritage for man. It says more about "who we think we are" than "where we came from."

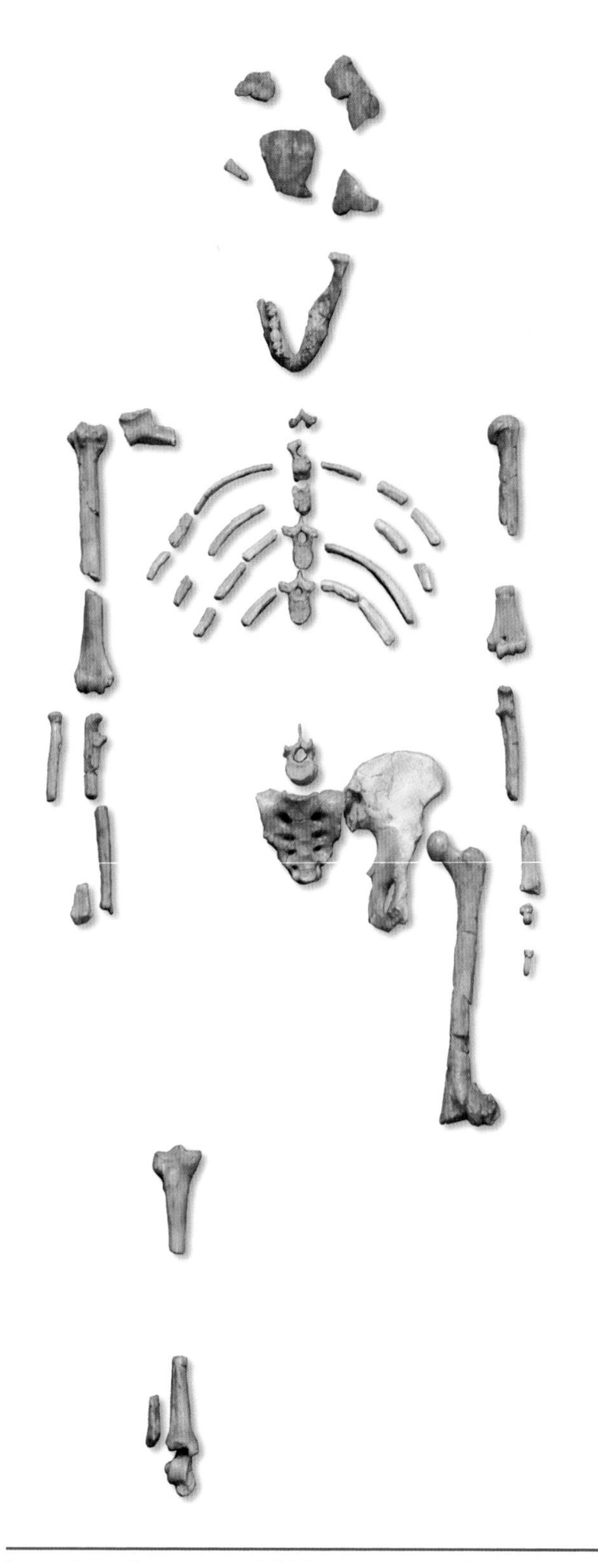

Lucy, long the poster child for human evolution. Only three feet six inches tall, with a chimp-size brain, chimp-like teeth and jaw, and long, curved fingers and toes, she was rather chimp-like. The only feature that pointed toward man was her knee, which was not found with the rest of the bones, and her pelvis. To some investigators, these features indicated that she could walk somewhat erect. The pelvis had been badly crushed, and when it was first reassembled it was substantially chimp-like. Only when manipulated artificially could it be presented as human-like. Could Lucy have actually been just a chimp?

true ape-man—is still missing. As the story goes, Lucy was able to walk upright, but could take to the trees when necessary. Is this accurate? Does this necessarily indicate that she was a human ancestor? Let's evaluate the evidence.

We note at once that she was quite chimp-like. The Lucy fossil was three feet six inches tall, a height typical of today's chimps. Furthermore, the Lucy fossil was completely chimp-like from the waist up. Her jaw was decidedly V-shaped, not U-shaped like a human's. The hyoid bone, important in breathing and vocalizing, was chimp-like. The shoulder blade resembled that of a juvenile chimp or gorilla, indicating that the creatures partially supported their weight by walking on their knuckles, with the spine kept rather horizontal. This shoulder blade positioned the arms such that they could be easily lifted over the head, as modern apes do while climbing in trees.

Similar fossils of other australopithecine specimens were found in the region. These had long, curved fingers and long, curved toes, with an opposable "thumb" on the feet necessary for swinging among the branches of trees. These more recently discovered bones of tree-climbing australopithecines were dated as coming after the time of the Lucy fossil. They should therefore have been more "evolved," but they were decidedly ape-like.

A comparison of the shape of australopithecine feet with the well-known footprints found at nearby Laetoli is also instructive. The Laetoli footprints are identical to footprints of modern humans and could not have been made by a creature with an opposable big toe. According to standard dating techniques, the two sites are deemed to be nearly the same age. Evolutionists insist they must have been made by the Lucy species, even though the footprints themselves are quite incompatible with the chimp-like foot of Lucy's australopithecine group. The human-like Laetoli prints were most likely made by a human.

Australopithecus had long, powerful forearms and short robust hind limbs, reminiscent of apes. Its ear structure indicated that it had a keen sense of balance, quite like a chimp and rather different from the canals in a human's ear. The teeth were also chimp-like, albeit with a few features similar to humans. It seems as though nearly every feature from the waist up was chimp-like, as well as many aspects of the legs.

Upright stance is taken as diagnostic of the human

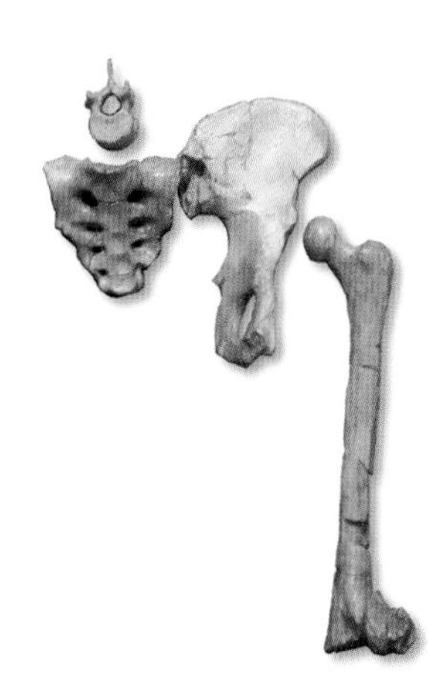

line. However, the well-known modern pygmy chimp, or bonobo, habitually walks rather upright, although somewhat differently from humans. No one claims that the pygmy chimp is more human than other chimps.

From the waist down, only two features of Lucy seemed to point toward an upright stance—the knee and the hip. The knee has engendered major questions related to its inclusion with the rest of Lucy. It had been found the previous year at a different location from the rest of Lucy's bones. Anatomical similarity between the specimens seemed to Johanson to warrant including it with Lucy, although others have disagreed. Even if it were included, however, professional anatomists have disagreed with Johanson's belief that it indicated upright walking in a human manner.

Lucy's Hip and Knee

The hip bone is the main indicator of stance and gait, and how the legs exit the hip and are oriented. Advocates make much of the configuration of Lucy' hip, but certain skeptics—evolutionists themselves—call attention to various discrepancies. Their contention is that australopithecines spent much of their time in the trees. While they may have been able to walk somewhat upright for brief periods, they did not walk as humans walked.

When Johanson formed his original conclusion, the hip he examined consisted of only one side of the pelvis. To date, no other adult *Australopithecus* pelvis has been found. More recently, the fairly complete body of an *Australopithecus* infant has emerged, and it is grossly ape-like. Chimp infants are known to undergo major body changes as they grow, so a definitive conclusion evades researchers. Thus, much of the entire contention of Lucy's upright stance is based on a partial hip.

Johanson drew several conclusions from his early studies, but over the years he and his colleagues have made numerous other claims. A documentary of his discoveries was produced in 1994 by the frequently pro-evolution Public Broadcasting System. It was a three-part NOVA series, *In Search of Human Origins*, featuring Johanson's discovery and preparation of Lucy's bones. Even though it is now several years old, it is instructive to revisit this program and quote a transcript of part of its dialog.

In episode one, "The Story of Lucy," Johanson relates that the pelvis as found initially appeared to be that of a chimpanzee and not from a creature capable of walking erect. The small pelvis bone had been crushed into about 40 pieces. He needed to accurately reconstruct it if Lucy's posture could be discerned. How could it be done?

Working with him was Dr. Owen Lovejoy of Kent State University, who "came to the rescue." Lovejoy first separated all the pieces from the surrounding matrix. He wanted to reassemble them in the proper order, but when he tried to work the 40-piece puzzle, the pieces did not fit in the hoped-for manner. Instead, when fit together, they looked like a chimp's pelvis. What follows is from the actual transcript of the documentary posted on the PBS website.

> DON JOHANSON: We needed Owen Lovejoy's expertise again, because the evidence wasn't quite adding up. The knee looked human, but the shape of her hip didn't. Superficially, her hip resembled a chimpanzee's, which meant that Lucy couldn't possibly have walked like a modern human. But Lovejoy noticed something odd about the way the bones had been fossilized.
>
> OWEN LOVEJOY: When I put the two parts of the pelvis together that we had, this part of the pelvis has pressed so hard and so completely into this one, that it caused it to be broken into a series of individual pieces, which were then fused together in later fossilization.
>
> DON JOHANSON: After Lucy died, some of her bones lying in the mud must have been crushed or broken, perhaps by animals browsing at the lake shore.
>
> OWEN LOVEJOY: This has caused the two bones in fact to fit together so well that they're in an anatomically impossible position.
>
> DON JOHANSON: The perfect fit was an il-

lusion that made Lucy's hip bones seems to flair out like a chimps. But all was not lost. Lovejoy decided he could restore the pelvis to its natural shape. He didn't want to tamper with the original, so he made a copy in plaster. He cut the damaged pieces out and put them back together the way they were before Lucy died. It was a tricky job, but after taking the kink out of the pelvis, it all fit together perfectly, like a three-dimensional jigsaw puzzle. As a result, the angle of the hip looks nothing like a chimps, but a lot like ours. Anatomically at least, Lucy could stand like a human.[28]

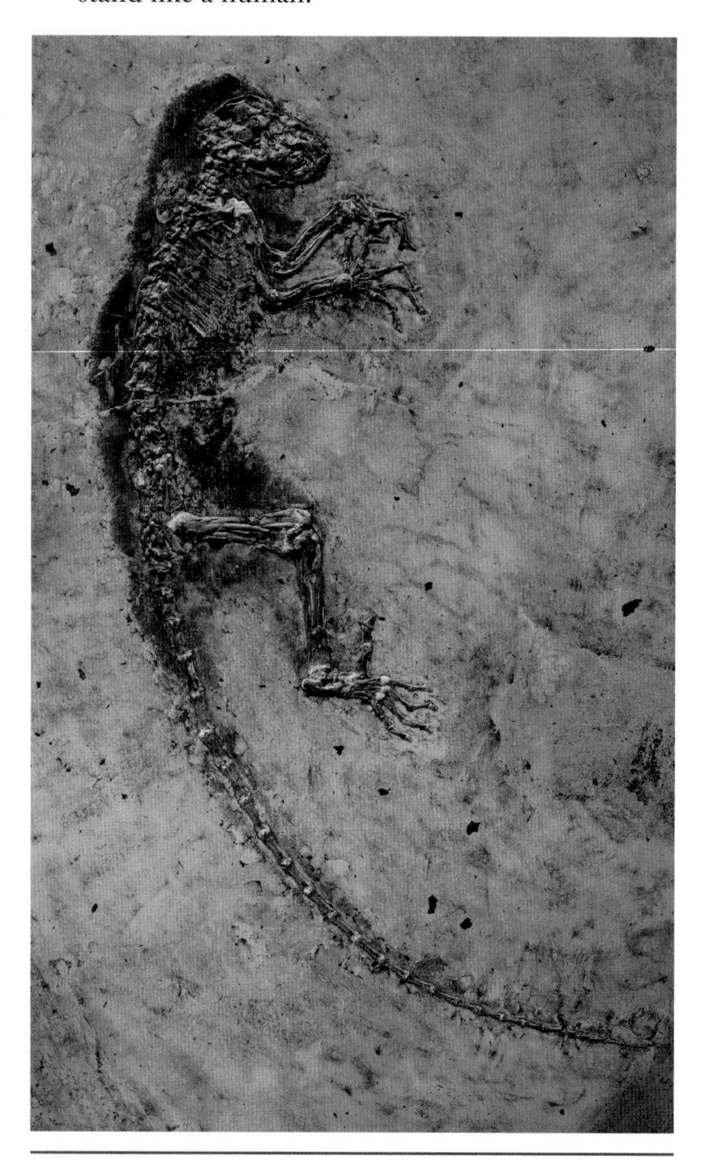

The year 2009 was the 200th anniversary of Darwin's birth and the 150th anniversary of the publication of his world-changing book. This occasion led to numerous attempts to promote evolution. "Ida," the fossil shown here, was hailed as the long-awaited link between man and animal. It soon became obvious that this creature was nothing more than a variety of lemur, and not on an evolutionary line to humans. Although gladly received at first, Ida was soon abandoned, even by evolutionists.

In the film, Lovejoy is seen using a grinding wheel to cut the pieces, shape them, and place them so that they mesh into a human-like form rather than the original ape-like form. He justifies this by using the analogy of a broken teacup, which can be glued back together by one who knows how. But this analogy presupposes that one knows that the fragments were, in fact, from a broken teacup. It presupposes that he knows the correct original shape of the teacup, that none of the pieces are missing, and that all the fragments being assembled are actually from a single teacup. His reconstruction of the shattered hipbone—the type of which had not been found elsewhere, and thus had an unknown original shape—has been the basis of years of evolutionary speculation and has spawned inferences on Lucy being "the first human" or the first step toward humanity. But what if the reconstruction, which was so intricate and extensive, resulted in unintended error?

Remember our two recommended questions for evaluating claims.

1. What do you mean by that assertion?

In this case, the claim is that fossils of a chimp-like animal—with many features of a chimp, chimp-like jaws and teeth, who swung from trees and had long, curved fingers and toes and a decidedly chimp-size brain—was the first member of the human line and could walk upright in the human manner. A host of related questions could follow.

2. On what evidence do you base that conclusion?

Is the evidence sufficient and convincing? In this case, we have a partial skeleton of a chimp-like creature that possessed knee and hip bones that under some circumstances could be viewed as similar to the human form. But the knee is perhaps disassociated from the rest, and the hip was artificially manipulated to make it appear less chimp-like and more human-like. Is this sufficient evidence on which to claim that man came from the animals, a claim that has such meaningful life implications?

This presentation and line of questioning is not meant to be mean-spirited or accusatory. It is not intended to imply that Johanson and Lovejoy fraudulently altered the data to force a desired conclusion. It is intended to help us think about such an "outrageous hypothesis" and to what extent an investigator's *a priori* convictions may have colored his thinking and work. It is intended to encourage critical thinking about the evidence used to support any historical view of the unseen past. Is there another interpretation of the data? Is there a better interpretation?

Even if Lucy could and did habitually walk upright, she was still a chimp. Perhaps she could walk more upright than most chimps, but that does not make her human or on the line to becoming human. There is a great difference between chimps and humans, and it has less to do with the body than with the soul and spirit. Materialistic evolutionists deny the existence of man's spiritual nature. They therefore have no option but to focus on the body. There is much variety among the apes and much variety among humans, but a vast gulf exists between the two groups that can never be bridged by anatomy alone.

Museums and textbooks that present Lucy as the beginning of human evolution often do so in a clever way. It usually takes the form of a diorama, showing not Lucy's bones, but a life-size model of Lucy as if she were alive. In several prominent American museums, Lucy is shown completely upright in a life pose, yet at only three feet six inches tall. She stands as a young woman, nude and anatomically complete, with a thoughtful look on her face. She may be a little too hairy for most people's taste, but what young person would not stand and stare at this naked girl? Or what adult, for that matter? What young person would not be impacted by the presentation, and how would they avoid the lasting impression that Lucy's kind evolved into our kind? Evolution claims we came from an animal like this.

Is evolution so desperate for believers that they will even use graphic pornography to get their message across? There must be a better message—one that is believable without trickery.

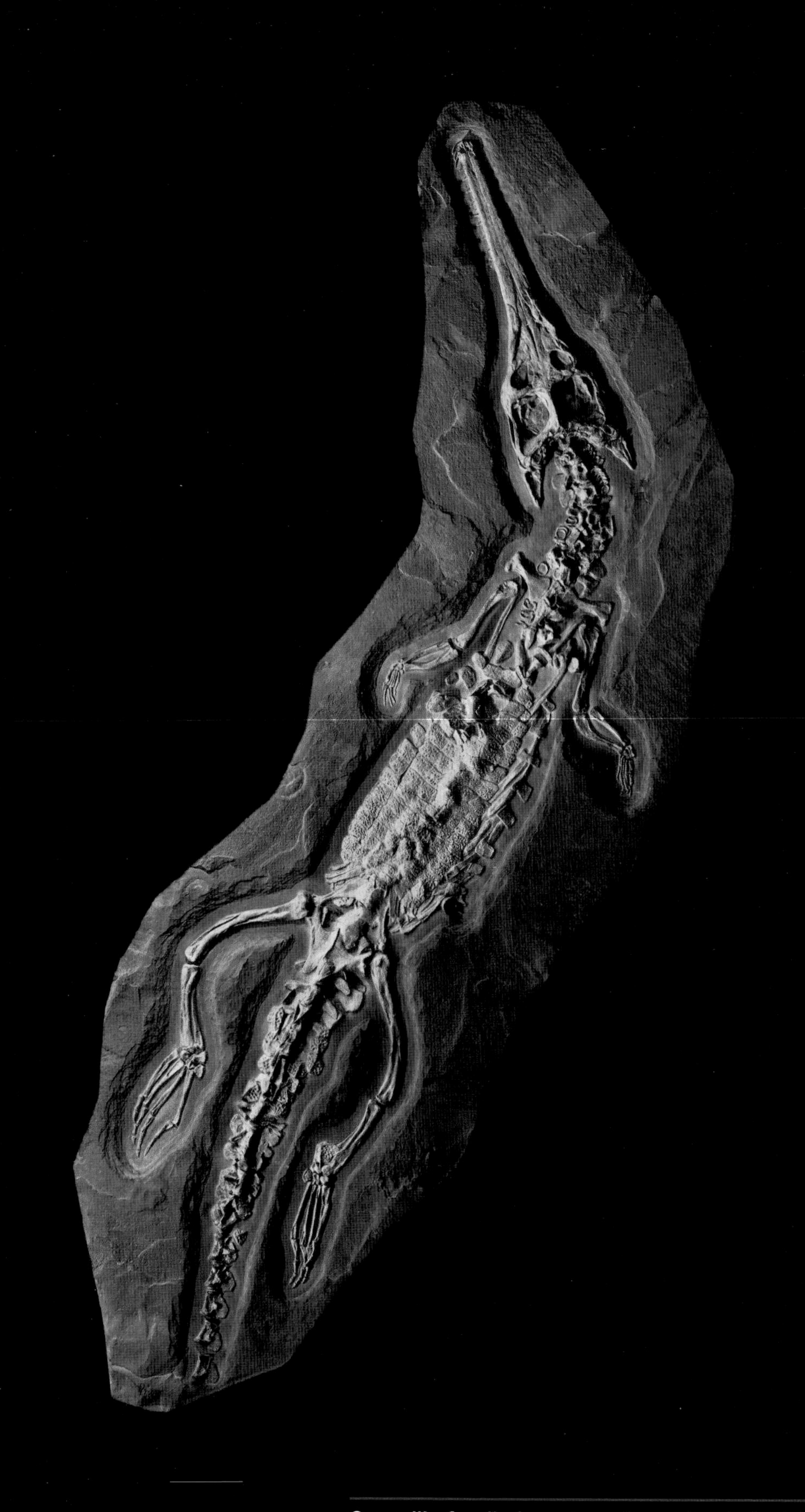

Crocodile fossils have been found in strata dated as "old as the dinosaurs." And they look just like modern crocodiles. Significant variety exists among living crocodiles, as it does among the fossils. This one sports a heavily armored body, much as crocodiles today that live in estuaries. (Location: Germany. Dated as Jurassic.)

Chapter 11 Fossils That Tell a Special Story

Evolution can be described in various ways, but at the very least it implies "survival of the fittest," the selection of those types that are best equipped to thrive and evolve. The corollary of this maxim is that "the unfit" do not survive. Their types go extinct. Over evolutionary time, a plant or animal type supposedly acquires helpful mutations and/or beneficial genetic recombinations and passes on these favorable changes to succeeding generations. Those without favored traits are unable to compete and die out. As it relates to fossils, a trend should be evident. There should be a visible change over time that is recorded in stone, with the more fit replacing the less fit, while the vast majority go extinct. There should be few (if any) exceptions to the evolutionary rule of "descent with modification."

Living Fossils

Yet there are exceptions. In fact, the overwhelming trend in the fossil record is one of stasis, with life forms exhibiting little or no change. The fossils are often of "persistent" types that do not appear to have changed much over the time their strata were being deposited. To make matters worse for evolution, sometimes a living animal or plant is found that had been thought to be extinct for "millions of years." It may even have been used as an "index fossil," a unique indicator of a particular stratum, and yet it had undergone no essential change from that time until the present. Darwin named these unusual discoveries "living fossils."

The most prominent example of a living fossil is the coelacanth fish already discussed, which was claimed as an important evolutionary "missing link." It was an index fossil thought to have evolved in the Upper Devonian period, after which it stayed much the same for about 200 million years before apparently becoming extinct. Now we discover, another 100 million years later, that they actually survived into modern times. Its form exhibited stasis in the extreme, and none of the extant specimens was at all like the transitional fish/amphibian that evolutionists had predicted.

With more extensive exploration, new cases of living fossils turn up often. An abbreviated list serves to make the point. (The times given are based on evolutionary dating.)

- Graptolites had been considered Ordovician index fossils, dated around 300 million years ago. Yet

The mussels below have been "opalized," infiltrated with silica until they have a sheen almost like a gemstone. They were crammed together in a mass burial, unable to free themselves from the gathering sediment. The horseshoe crab (right) bears an eerie similarity to the extinct trilobite. (Location: Australia. Dated as lower Cretaceous.)

rect. For instance, many people have been told that it takes long ages to petrify wood. But numerous examples of rapidly petrified wood have been documented. Often the timeframe can be known, because the wood was from an artifact of recent human activity.

Consider a specimen found in the state of Washington. This area is replete with volcanic ash, which is perhaps the best environment in which to petrify wood. Water percolating through the ash may carry dissolved silica, especially if it is hot. The water seeps into the wood, emplacing silica by surrounding, filling in, or replacing each cell, thus transforming it into what is essentially a rock in the shape of the original wood.

This particular piece of petrified wood came from a fencepost that was installed sometime in the mid-1800s and discovered around the year 2000. The part of the post above ground had rotted away, but the underground portion petrified. There are many such examples of rapid petrification, especially in volcanically active regions.

Recently, a team studied the degree of petrification of naturally occurring wood fragments that had fallen into hot, silica-rich waters in Japan. They also immersed freshly cut timber in the pool. When they checked the degree of petrification after seven years, they found that the wood was 40 percent petrified, or silica saturated.[3]

The petrification was accomplished as silica spheres, much like tiny beads of opal, precipitated from the water. Neither the petrification of wood nor the formation of opal takes an excessively long time—it just takes the right conditions. Remember that the great Flood was accompanied by extensive volcanism and voluminous devastation of forests.

Petrified Forest, Arizona

A deposit of petrified wood lies in northern Arizona adjacent to the Painted Desert. Here, abundant petrified wood can be found, both in and outside the park, providing souvenirs for rock shops around the world. Within the laterally extensive Chinle Formation, beautifully colored petrified wood can be seen, often with the tree rings intact. As to its origin, visitors are told

It does not take long to petrify wood. It just takes the right conditions—mainly water, preferably hot water saturated with dissolved silica. Such situations are common in volcanic regions. Sometimes, even the tree rings are preserved. The petrified wood on the right is actually the end of a split rail fencepost, partially buried in volcanic ash in Washington state in the mid-1800s. While the upper post rotted away, the portion underground petrified.

A polystrate fossil is a fossil that extends through many (poly) layers of rock (strata). This fossilized tree in Tennessee (one of many in the area) pierces numerous layers of shale and coal, which were thought to require long ages of time to be deposited in a marine environment. But the once-mature tree grew on land. Even if it had been somehow submerged, it could not long have survived underwater. The tree must have been uprooted and floated to its place of burial, maintaining an upright orientation as the strata accumulated rapidly around it. The Petrified Forest of Yellowstone National Park has long been thought to "disprove" the Bible's depiction of a young earth. However, the discovery of matching tree ring patterns in trees from multiple layers, showing that they grew at the same time, put an end to that. The knowledge that the upright trees penetrate more than one stratigraphic layer, that the root ball of one overlaps that of the other, and that the trees were rafted in by mudflows and did not grow in the place where they are now found, confirms their "polystrate" status.

Charles Lyell wrongly used the polystrate trees of the classic Joggins site in Canada to promote uniformitarianism and great ages of geologic time. He claimed that they grew where they are found, but now we know that they extend through several strata layers, including thin coal layers, disproving that claim. The trees were transported by water to this location and deposited in rapidly gathering sediments. They grew in pre-Flood days, but are dated as Pennsylvanian.

stories of a prehistoric standing forest that was overwhelmed by a nearby volcanic eruption and ash fall, preserving the trees in their place of growth. Is this the correct interpretation?

Several observations have been ignored to arrive at this "story." All petrified trees in this location are prone, having been knocked over rather than having merely grown old and fallen over. Missing are the root systems of the trees and all the bark. This seems strange if the trees are simply lying where they fell. None of the trees are at present standing upright in place of growth.

Petrified remains of upland conifer trees are the most common of the fossils. Closer inspection reveals a great variety of plant and animal fossils from a wide variety of habitats, from fern fossils to deciduous trees. Less common, but present, are large reptiles and dragonflies. This was not a once-standing forest. Despite the variety, no complete ecosystem is present. Instead, mixed environments are represented, which is not what would be expected if a forest was buried by a sudden volcanic eruption. Furthermore, the trees exhibit a preferred orientation, as if transported by water or a flowing mud. After lengthy study, Dr. R. C. Moore disavowed the standing forest "story" in favor of a regional cataclysmic event. He claimed:

> There lie thousands of fossilized logs, many of them broken up into short segments, others complete and unbroken....The average diameter of the logs is 3 to 4 feet, and the length 60-80 feet. Some logs 7 feet in greatest diameter and 125 feet long have been observed. None are standing in position of growth but, with branches stripped, lie scattered about as though floated by running water until stranded and subsequently buried in the places where they are now found. The original forests may have been scores of miles distant. The cell structure and fibers have been almost perfectly preserved by molecular replacement of silica.[4]

Polystrate Trees

In the Petrified Forest of Yellowstone National Park, upright trees are occasionally observed to penetrate through several layers of the overlying mudflows. If hundreds of years or more passed between depositional events, this could not be, since a tree in the lower layer would have died and rotted away before the next layer could encase it. These are called polystrate ("many strata") fossils. Such fossils are well-known but seldom discussed by geologists. Fossils that transgress more than one so-called "time horizon" cannot represent a long sequence of events, as is commonly taught.

Sometimes the body of a fossilized animal straddles many "years" of deposition. One can only conclude that the deposition was rapid and that the fossil was buried catastrophically in an area that was receiving a continual supply of sediments that built up around the recently deceased—or still alive—specimen.

Perhaps the most obvious polystrates are found in coal regions, where supposedly terrestrial swamp deposits are found intermingled with marine sediments. Often, upright stumps can be seen penetrating through several layers of coal, shale, siltstone, and limestone, all of which are thought to take a long time to accumulate and require totally different environments of deposition.

Human Fossils

How do human fossils fit in the biblical picture? If most of the fossil record resulted from the Flood, and the Flood destroyed all of humanity except for the individuals on the Ark, why don't we find more human fossils?

As mentioned earlier, land creatures have a "low-fossilization potential." They form a very small percentage of the fossil record, which is what is predicted by the creation model. Land-based creatures would have been the final victims of the Flood as it inundated the continents. Their remains would have been subjected to forces and processes that would make preservation highly improbable. In a dynamic worldwide flood, neither humans nor mammals in any great numbers would likely be preserved.

Evolution, however, holds that humans have lived on earth for millions of years. Surely there would have been at least as many opportunities for their fossilization as for other creatures. Why are so few ancient human remains found, if evolution is true? The fossil record, once again, speaks to the creation model.

Living Fossils

Here are just a few examples of creatures that appear in the fossil record in the same forms we see in their modern counterparts. Examples come from different supposed geologic time periods, but the message is the same: Evolution did not happen.

Beetle

Dragonfly

Hemiptera

Lobster

Pterodactylus kochi. Due to the delicate lightweight construction of their hollow-boned skeletons, a charac-
eristic that was essential to their flying abilities, these reptiles are rarely preserved in the fossil record. This
spectacular specimen represents one of the few finds of a well-articulated *Pterodactylus*. It features an excellen
skull, 3¼ inches in length, boasting many sharp and pointed teeth. The quality of the preservation is so fine tha
one can even distinguish many of the tiny claws on the feet. These flying reptiles are often wrongly classed with
the dinosaurs. They do not merit that designation, however, for they did not have the right hip structure. It was
neither a "bird hip" nor a "lizard hip," and was not remotely like the hip of a bird. The long, robust fourth finge
supported the leathery wing membrane. No animal ever discovered fits the requirements of a possible ancesto

Chapter 12 The Real Nature of the Fossil Record

The fossil record leaves an inescapable impression on the honest observer. It certainly does not communicate the macroevolutionary picture. The record of the past written in stone contains no evidence that any particular animal ever morphed into a fundamentally different type of animal. No trend can be found of gradual, Darwinian alteration through mutation and natural selection. These processes occur, but they are not mechanisms for true evolution of basic body styles.

Nor do we see punctuated equilibrium transforming them rapidly. Without a doubt, we see sudden changes in dominant fossil shapes as we ascend the geologic column, but this is not macroevolution. The species changes touted by punctuated equilibrium that we do see are either common variation of individual offspring, or adaptation of a population to differing conditions. Punctuated equilibrium does not even address the larger changes needed for meaningful evolution.

On the other hand, the fossil record does communicate the sudden appearance of basic types, complete with all the features that characterize them. Lots of variety is on display, even at times enough to lead to a new species. But variety is not evolution. Cats are cats and dogs are dogs and always have been so. There are similarities between them but no hint of relatedness. Both appear to have been suddenly created to live in similar environments, breathe the same air, eat the same foods, drink the same water, and survive through circulation of similar blood. We should expect similarities. But cats when they reproduce yield kittens, and reproducing dogs have puppies. They did not originate by mutations in a different type of common ancestor, nor did one come from the other. And this is what the fossils show.

Once a basic type appeared, it demonstrated stasis. Individuals varied in appearance and whole populations varied over the generations to accommodate changing conditions as they "multiplied and filled" earth's varied environments, but always they were fundamentally the same as the parent group. The fossil record features stasis as a dominant trend. It does not speak of major changes. Evolution, or the descent from a common ancestor model, demands that major changes visited every population. But this is the evolution "story," not the conclusion drawn from the fossils.

Each plant or animal alive today exhibits amazing complexity at the start. Each of its body parts is precisely designed to perform its function, and all work together for the good of the whole. Indeed, there often is no use for a particular part without the others. Some may only be used at a limited period of life or in an unusual circumstance, at which time they must be present for survival. All must be present for any to accomplish any useful purpose. From the very first time a fossil type appears (i.e., the lowest stratigraphic interval in which it is found), it shows all the design features which make it special. Evolution necessitates the gradual accumulation of body parts through random mutation and the amalgamation of previous parts with different functions into a new whole. The elegance of design, however, argues *against* a patchwork origin and *for* an intelligent cause. Mutations only mar, but do not erase, evidence of exquisite design.

Extinction is well-documented in the fossil record, and while extinction is a necessary part of the evolutionary scenario, it is not evolution. It might better be considered as the opposite of evolution. Losing a type is not what is in question, but the gaining of new types—now, that would be interesting. A case can be made by fossil "splitters" that new species can be found as one ascends the strata. However, speciation within basic kinds is different from the introduction of new kinds, and evolution requires a dizzying array of basic new kinds. The origination of a new form has never been documented in the modern world of scientific observation, while perhaps several species every day go extinct. The opposite of evolution occurs today, and fossils show that the opposite of evolution also occurred in the past.

Evolution necessarily implies the concept of "descent from a common ancestor or ancestors." Yet no ancestor/descendant relationship can be advocated with certainty based on the fossils. Indeed, the differences are obvious and make classification between types possible. The similarities between distinct types is not a sure footing on which to base an ancestral relationship, as proved by the many mutually exclusive cladograms advanced by evolutionists. Whose opinion, if any, is correct? The separateness of each type is witness to their separate creations.

The fossil record can be deemed essentially complete. Darwin was concerned about its lack of transitional forms, hypothetical creatures that demonstrate one type changing into another over time. He was hopeful they would be found one day. But extensive exploration and fossil discovery in the following years have not brought such in-between forms to light. The vast majority of taxonomic orders and families that live today are also found as fossils, yet without fossil transitions. We can be certain the record is substantially complete.

The "Cambrian explosion" constitutes a major episode in the history of life. If evolution were true, one would expect the record to start with one type of animal life, then increase to two, and so on. Yet fossil studies have shown that essentially all phyla were present at the start, each distinct from the others and each fully equipped to function and survive. Even vertebrate fish were present in the lower Cambrian. Some phyla have gone extinct over subsequent years, but most have continued into the present. There is no evolutionary tree found in the fossils as Darwin and his disciples have claimed. Rather, it is more like a lawn than a tree.

Stasis can be seen in the large, vertical, stratigraphic ranges of many fossil types. Index fossils are thought to exist only for a brief time span. True enough, some fossils are only found in a relatively few layers. But many other fossils, such as the brachiopod *Lingula*, can be found throughout the geologic column and into the present. This animal would seemingly make a good potential ancestor for others, but it never changed into anything nor arose from anything. Various fossil types are found in many layers, with more fossil ranges being continually extended by new discoveries. Statistical treatments give reason to believe that essentially all types lived throughout a large portion of history.

At least 95 percent of all animal fossils are of marine invertebrates. They are found in great variety, but are well-designed for life in the sea. Some lived in high-energy, near-shore environments, but others lived in the deep ocean, away from the pounding action of the waves. Among the vertebrates, most fossils are fish, again mostly marine creatures. Of the terrestrial fossils, by far most are plants. Land-dwelling animals, such as mammals and dinosaurs, are poorly represented in the fossil record. The majority of animals depicted on evolutionary fossil charts in textbooks, however, are land vertebrates. It is thought that a possible case for evolution can be made from them, but this does not accurately portray the real fossil record.

These fossilized marine creatures are typically found in catastrophic deposits. Even marine creatures that live in high-energy zones cannot live in catastrophic conditions. Many died where they were fossilized. They were either buried alive, or their remains were transported by dynamic processes to their present resting places before they could decay or be scavenged. The processes involved must have been highly destructive, yet rapidly acting. Major forces that sculpted the earth's surface were not like processes of today.

Often the fossil remains are found in a death pose. The famous *Archaeopteryx* fossil lies with its neck and tail arched back as if it were dying a horrible, drowning death. Clams are found with both halves tightly shut, "clammed up" as a living clam does for protection from danger. Dinosaur fossils, also in death poses, are found ripped apart but often not scavenged. Fossilized animals give every indication they were violently killed and/or transported to the places we now find them.

The fossils are usually entombed in deposits with no complete ecosystems present that could have supported them in life. Often evolutionists portray a fossil's tomb as a snapshot of life and tell stories about the creature's habits. But these plants and animals are not necessarily found where they lived, or where they died. They are found where they were buried. It is not honest to presume patterns of life from transported remains of once-living things.

Fossil graveyards often contain numerous animals from mixed habitats. Saltwater fish are sometimes found with upland dwellers. Crocodile fossils are found with deep sea denizens and desert and arctic mammals. They could scarcely be lumped together in this way by the uniform processes of today. Some great cataclysm is needed.

The catastrophic deposits in which the (mostly marine) fossils are found are almost all on the continents. A series of marine cataclysms inundated the land, destroying nearly everything there and laying down a record plain enough for all to see. Those terrestrial fossils that were deposited primarily date from the Ice Age that followed the great Flood of Noah's day.

Summary

Combining all these major concepts, we see that the fossil record is a record quite different from that presented in support of evolution. Each basic plant and animal type appeared abruptly and fully functional, and then experienced stasis throughout its tenure. Each type was complex and distinct at the start, without having descended from some other ancestral type, particularly from a less complex type. All basic types that have ever lived were present at the start, and while some have subsequently gone extinct, no new basic types have appeared since the beginning. We have reason to believe that substantially all basic forms that ever lived have been found as fossils.

A general rule is that the fossils extend through a lengthy stratigraphic range with little or no change. Most of the fossils are remains of marine invertebrates, found in catastrophic deposits, often in death poses with an incomplete ecosystem present. These predominately marine fossils are almost all found on the continents, not in the ocean.

The fossil record is thus quite incompatible with evolution and uniformitarianism, but remarkably consistent with the biblical record: Creation of all things in perfect form and function, the curse on all things due to man's rebellion, and the great Flood of Noah's day that first destroyed and then renovated the entire planet.

Creation thinking predicts the evidence, while evolution must distort and flex the evidence and its position to accommodate it.

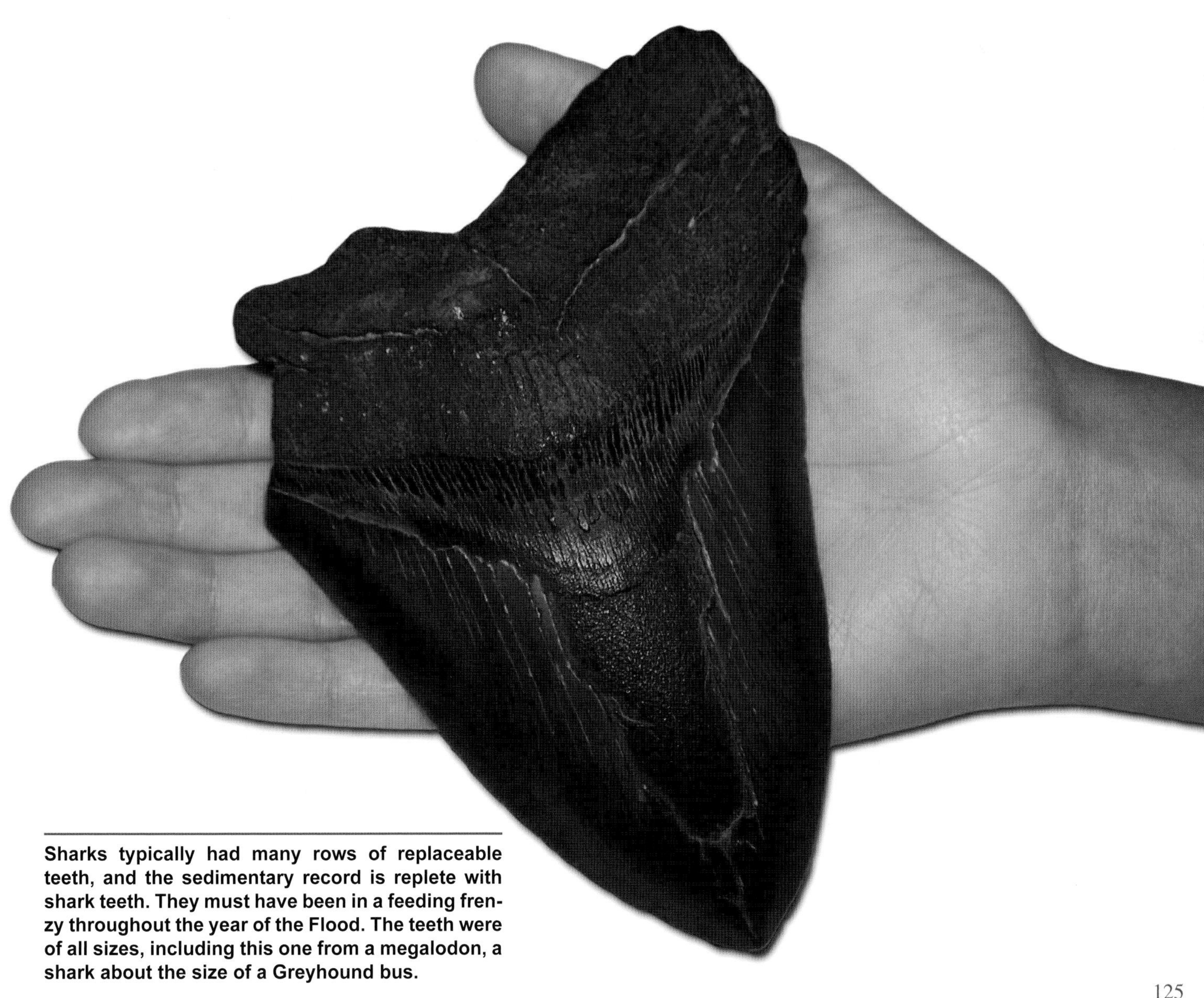

Sharks typically had many rows of replaceable teeth, and the sedimentary record is replete with shark teeth. They must have been in a feeding frenzy throughout the year of the Flood. The teeth were of all sizes, including this one from a megalodon, a shark about the size of a Greyhound bus.

Chapter 13 What Difference Does It Make to Me?

What we have seen from a study of the fossils convinces us that evolution did not occur in the past. The record testifies to a different history and thus to a different origin of plant and animal types. The record speaks of the abrupt appearance of each type, fully formed and fully functional from the start. Each type possessed a marvelous complexity and design from the beginning. Things did not arrive by the steady transformation of previously existing types through random mutation and natural selection. Evidence of the descent from a common ancestor is systematically lacking. Yet life exists today in great profusion and precise design. Life must have originated through some process other than random mutation.

The fossils record exhibits not only abrupt appearance and sudden complexity, but a pervasive stasis. Types stubbornly refuse to change outside of narrow limits. Gaps that separate every basic type from every other type are regular and systematic, mirroring those in living types. Abundant types were present in the beginning. No basic types have originated since that time, though some have gone extinct.

The fossils also speak eloquently of their lifetimes having been dominated by death and disease. Many lived a miserable life and met a tragic end. Large graveyards full of fossils, quite unlike death zones today, dominate the strata.

We are able in some instances to reconstruct the cataclysmic events that buried the fossils, and often find that widespread flooding was responsible. Fossils are strewn about, with bodies broken and dismembered, together with fossils from varying habitats. Geologic studies likewise testify to catastrophic forces. What could have been the cause?

Scripture, which claims to be the written account of the Creator Himself, reveals a past time when God separately created each basic plant and animal type. These complex creations all played a vital role in earth's early environment. According to His Word, creation was perfect at the start, but suffered ruination when the head steward of creation rejected God's authority over him. Soon a great worldwide flood cleansed the earth of its rebellious human inhabitants, and all the earth suffered. Today we study the remains of that once "very good" world, devastated by man's rebellion.

Thus, the fossils record the same message given in writing by the Creator. His message is totally trustworthy, for He was there, knows everything, can communicate clearly, and cannot lie. He wants us to know, and realizes we might come to error if we try to unravel past secrets alone.

Today, the human manufacturer of a car has the knowledge to explain to an owner how and when the car was constructed, and indeed has the responsibility to explain how such knowledge might impact the owner's use of the car. It would be presumptuous to dispute the manufacturer's claims for design and time of origin. Since we were not there, we cannot fully know the details unless we are told. We can rightly discern that the car was carefully designed, and see that nothing was left to chance, but the details of how and when are beyond our ability to discover with certainty.

One would also do well to heed the manufacturer's instructions to use regular gas in a car with an internal combustion engine instead of diesel gas. To disobey is to invite disaster. In a similar fashion, our Creator has the authority and knowledge to tell us how we should live if we are to function properly. He has the authority to set the guidelines for life and the penalty for disobeying those guidelines. If we live according to His "owner's manual," we can expect to enjoy life more fully than if we choose otherwise.

Fossils and the other historical sciences tell us a lot, but they do not provide all the details. An accurate reconstruction of past events from partial evidence in the present is nearly impossible. Ask anyone involved in forensic science. We can never be sure of the events that took place. In just that way, studying the fossils provides only clues in our quest to understand the past. We do not know all truth. We can only come to a better approximation of that truth as we do our investigating under the Creator's authority. Never can we discover truth if we have already denied truth. But with His help, we can begin to fill in the blanks

The fossil record speaks eloquently of the Creator's precise design and of the truth of the Bible's message to us. There we are told of His creative power, His judgment of sinful rebellion, and His ultimate solution to the sin problem. Before Adam sinned, God justly declared the penalty of sin to be death, eternal death. Yet He desired a relationship with sinful man. Knowing that man was powerless to overcome this sentence on his own, God moved to satisfy it Himself. This was accomplished by the death of the God's only Son, Himself the agent of Creation, in payment for our sins. As a sinless man, Jesus Christ could die for another's sins. As the Creator God, His death could apply to all. Only in this way could God's holy and just nature be satisfied, and man brought back into right standing with his Creator.

God's plan of salvation is a free gift to any who confess their sin, turn from it, and accept His offer of rescue. He will respond with forgiveness of our sins, restoration of our standing before Him, adoption into His family, and eternal life in His presence. Such forgiveness can only be received by faith. Evidence, such as that preserved in the fossil record, can be a stimulus to faith, but evidence can never substitute for faith.

The authors trust that this book will be of assistance in your own quest for truth and an impetus to your growing faith. By way of testimony, we have both fully acknowledged His truthfulness and accepted His forgiveness, and attempt to live our lives according to His design. Doing so, we have found His way a joy, in both our personal lives and our professional careers. His truth informs our study and makes good science possible. It does not guarantee success in every venture, but allows us to do better science. We recommend Him and His truth to you.

APPENDIX

The True Nature of Evolution's "Transitional Forms"

Neo-Darwinists (evolutionists) look to the fossil record for support of their "molecules-to-man" theory. They have been plagued, however, with a critical lack of transitions, a fact that non-Darwinists have fittingly exposed and promoted. If the strange evolutionary process were true as described by secular science, the media, and the educational system, then an enormous number of transitional fossils should be the norm. Indeed, considering the hundreds of millions of years of gradual evolution claimed by evolutionists, most of the fossil record should be comprised of "missing links."

This is not what is found. Still, evolutionary rebuttals are routinely attempted to counter the solid arguments presented by non-Darwinists that there are no links between the major groups of living creatures. Over 20 years ago, University of Washington biologist Kathleen Hunt put together a long, highly questionable Internet list of what she felt were transitional vertebrate forms. Many evolutionists have taken advantage of Ms. Hunt's extensive catalog and have copied her material in various blogs, forums, classes, and anti-creation letters.

Because Ms. Hunt's list is so extensive, and because it has been cited so widely, it offers a useful platform from which to consider the subject. Using quotations from major evolutionary textbooks, science dictionaries, and technical articles, as well as other sources, this appendix will examine some of those supposed "links" as a means of presenting the true nature of evolution's "transitional forms."

Introduction

Evolutionist Fred Guterl wrote an article in *Newsweek* claiming that Charles Darwin "had plenty of fossil evidence to back him up."[1] This would have been a surprise to Darwin, who included an interesting chapter in his 1859 book *On the Origin of Species* entitled "On the Imperfection of the Geological Record." In this chapter, Darwin spoke of "the sudden appearance" of major fossil groups in the Cambrian, one of the lowest fossil-bearing strata.

Geologist Donald Prothero of Occidental College wrote a piece for *New Scientist*, stating that a "favourite lie" of creationists is 'there are no transitional fossils.'"[2] But according to paleontologist Michael Benton:

> A key question paleontologists always ask is whether the fossil record is good enough to tell the history of life or whether it is hopelessly riddled with error. Opinions have wavered back and forth over the years.[3]

Other secular writers admit to the absence of transitional evolutionary evidence.

> The fossil record is a unique source of evidence for important evolutionary phenomena such as transitions between major clades [a lineage branch in an evolutionary diagram]. Frustratingly, relevant fossils are still comparatively rare, most transitions have yet to be documented in detail and the mechanisms that underpin such events, typified by rapid large scale changes and for which microevolutionary processes seem insufficient, are still unclear.[4]
>
> After an early period of enthusiasm when fossils were thought to explain all of evolution we are now passing through a time of disillusion and fossils are regarded as having poor explanatory power as to the origin of major structural characteristics.[5]

Kathleen Hunt makes an incredible admission at the beginning of her list: "Few or none of the speciation events are preserved."[6] This is a very significant statement because she is admitting that where there *should* be a transition between two unique kinds of creatures—there are none. (Conversely, the creation model predicts there should be no unquestionable missing links.) The non-Darwinist would then have to ask just what is Ms. Hunt attempting to do with her Transitional Vertebrate Fossils FAQ if "few or none of the speciation events are preserved"?

Additionally, in her introduction Ms. Hunt admits to "major breaks" in the fossil record:

> For instance, the Aalenian (mid-Jurassic) has shown no known tetrapod fossils anywhere in the world, and other stratigraphic stages in the Carboniferous, Jurassic, and Cretaceous have produced only a few mangled tetrapods....To further complicate the picture, certain types of animals tend not to get fossilized—terrestrial animals, small animals, fragile animals, and forest-dwellers are worst. And finally, fossils from very early times just don't survive the passage of eons very well, what with all the folding, crushing, and melting that goes on. Due to these facts of life and death, there will always be some major breaks in the fossil record.

This is certainly true, and these major breaks just happen to be where creation science predicts they should appear—between the major kinds of vertebrates. Creationists do not have to apologize or make excuses regarding these clear breaks. Indeed, creationists can conclude, "Because of the fossil record: creation," while the evolutionist must say, "In spite of the fossil record: evolution."

Paradoxically, the supposed transitional forms Ms. Hunt does list are not embraced by most secular scientists. For example, the evolutionary authors listed in this appendix should have mentioned at least some—if not all—of the vertebrates listed in Ms. Hunt's FAQ with considerable fanfare. Indeed, entire chapters or sections in vertebrate paleontology textbooks should be written to list the enormous number of transitional forms.

What is found instead is a strange silence or at best diffidence in science texts and peer-reviewed papers. Kathleen Hunt continues her introduction by cryptically stating, "The second reason for gaps is that most fossils undoubtedly have not been found." Ms. Hunt evidently feels these undiscovered vertebrate fossils will eventually be found, whereas creationists maintain that the missing links will continue to be...missing.

Ms. Hunt states that the non-Darwinist's "complaint" that "there aren't any transitional fossils...has long been obsolete and inaccurate, as a brief glance at the fossil record shows." Very well, let us look to the fossil record and see what it reveals.

To begin with, if macroevolution were a fact it would be clearly documented in the vertebrate paleontology texts. Instead, what the reader regularly finds are phylogenetic trees or diagrams with dotted lines (i.e., no fossils found, but only predicted by evolutionary theory) at critical junctures. Some evolutionary texts are not as honest, placing deceptively solid lines between major vertebrate groups—clearly indicating evidence for macroevolution *where there is none*. By insisting the fossil record contains a multitude of "missing links" developed over millions of years, evolutionists reveal their deep misunderstanding of science. A professor of paleobiology stated:

> The known fossil record is not, and never has been, in accord with gradualism. What is remarkable is that, through a variety of historical circumstances, even the history of opposition has been obscured. Few modern paleontologists seem to have recognized that in the past century, as the biological historian William Coleman has recently written, "The majority of paleontologists felt their evidence simply contradicted Darwin's stress on minute, slow, and cumulative changes leading to species transformation."[7]

In this appendix, various major vertebrate groups and subgroups of living organisms will be briefly described and addressed, interspersed with expositions in the order of the major transitional categories listed by Kathleen Hunt in her Transitional Vertebrate Fossils FAQ. An introduction will be presented (with definitions as required) to each major group. After the introduction, the transitional form statements made by Ms. Hunt will be given. The answer to these supposed transitional forms will then be listed in three parts:

1. What the fossil evidence actually shows with no editorial comment from either side of the

origins issue.

2. The evolutionary interpretation of this fossil. (Kathleen Hunt's comment will sometimes be the only remark listed.)

3. The non-Darwinian interpretation—the fossil is not a transition.

The following evolutionary publications and references will be used throughout this appendix:

The Concise Dictionary of Zoology, 4th ed., edited by M. Allaby
Henderson's Dictionary of Biology, 15th ed., edited by E. Lawrence
The Penguin Dictionary of Biology, M. Thain and M. Hickman
Major Transitions in Vertebrate Evolution, edited by J. S. Anderson and H. Sues
Vertebrate Paleontology, 4th ed., M. J. Benton
Gaining Ground: The Origin and Evolution of Tetrapods, 2nd ed., J. A. Clack
The Origin and Evolution of Mammals, T. S. Kemp
Colbert's Evolution of the Vertebrates, E. H. Colbert, M. Morales, and E. C. Minkoff
Integrated Principles of Zoology, 15th ed., C. P. Hickman et al
Zoology, S. Miller and J. Harley
Vertebrate History: Problems in Evolution, B. Stahl
Early Vertebrates, P. Janvier
Vertebrates, 6th ed., K. Kardong
Evolution of the Rodents, P. G. Cox and L. Hautier
Great Transformations in Vertebrate Evolution, K. P. Dial, N. Shubin, and E. Brainerd
The Rise of Fishes, 2nd ed., J. Long

Chordates

Chordates may be defined as a large phylum containing animals that possess a notochord (or a rod of vacuolated cells) lying ventral to the neural tube at some stage of their development. They also have pharyngeal pouches and an endostyle—a shallow groove in the pharynx (throat) floor of protochordates that may develop as the thyroid gland in chordates. It is significant that the first chordates and "early" vertebrates have been found in Cambrian sediments, which is earlier than evolutionary theory says they should appear.

In 1974, evolutionist Barbara Stahl described the evolution of the early chordates in a purely imaginary ("must have been") manner:

> According to this theory, the animals at the base of the chordate line must have been soft-bodied, sessile forms of small size, resembling perhaps the minute pterobranch relatives of *Balanoglossus*, the acorn worm. These hypothetical forerunners of the chordates may have trapped food particles (as pterobranchs still do) by waving soft, unjointed appendages in the surrounding water.[8]

Six evolutionists maintain most zoologists must use imagination to "see" the chordate ancestor:

> Thus, most zoologists currently reject Garstang's hypothesis and *envision* the ancestral chordate as a free-swimming creature *perhaps* similar to the modern amphioxus.[9]

In 2015, Michael Benton stated:

> Urochordates [sea squirts, tunicates, phylum Chordata] have a patchy fossil record....The fossil record of cephalochordates is not much better.[10]

A newly named unusual group, the Vetulicolia, from the Chengjiang Formation in China, is "still highly controversial," according to Benton:

> Additional invertebrate chordates from Chengjiang, the yunnanozoons *Yunnanozoon* and *Haikouella* have been interpreted by rival researchers both upwards and downwards in the scheme of things. One team identified these animals first as possible cephalochordates, and then upwards as vertebrates. The other team preferred to regard the yunnanozoons first as hemichordates, and then downwards as basal deuterostomes allied to the vetulicolians. The problems revolve around different interpretations of coloured blobs, lines and squiggles in the fossils. There are plenty of fossils—literally thousands—but anatomical interpretation is critical.[11]

Other evolutionists conclude:

> But there is still much debate about the relationships between the animal groups that made their apparation [*sic*] very early in evolutionary history, probably in the late Precambrian, some 650 to 540 million years ago.[12]

Secular researchers are already (*a priori*) committed to belief in evolution, although there are no specimens living or fossilized that would even suggest a support of Darwin's theory of descent-with-modification (vertical evolution). Indeed, the same criteria apply to identifying both extant (living) and extinct members of the same group—just as a creationist would expect.

Vertebrates

Vertebrates may be defined as a chordate phylum (now termed Craniata) containing mammals, fish, amphibians, birds, and reptiles. They have a distinct head with a brain that is protected by a skull. Evolutionists have found it difficult to fit the development of vertebrates into an evolutionary picture.

> Zoologists do not know what animals were the first vertebrates.[13]
>
> Despite their critical importance for understanding the origins of vertebrates, phylogenetic studies of chordate relationships have provided equivocal results.[14]

Subdivision of Vertebrates

Amniotes are a nontaxonomic designation of a type of development for a group of "higher vertebrates" that are distinguished by the presence of extraembryonic membranes during embryonic development.

> Even though fossil records of many amniotes are abundant, much remains to be learned of amniote origins.[15]

Benton addresses relationships of early amniotes (i.e., turtles) by saying:

> Unexpectedly, micro-RNAs indicate a definite pairing of turtles and lepidosauromorphs (Lyson, *et al.*, 2012), a difficult issue to resolve at present. There is no molecular evidence that pairs Synapsida and Diapsida, with turtles as outgroup, as would have been expected from the traditional morphological phylogeny. [16]

The three groups of amniotes can be described thusly:

1. Anapsids (no temporal opening)—class Reptilia, tortoises and turtles.
2. Diapsids (two temporal openings)—birds, dinosaurs, lizards, crocodiles and snakes. Evolutionists say that the evolutionary history of the diapsid lineage is quite complex. Indeed, the term "diapsid" is dated, as evolutionists see the more "primitive" and more "advanced" forms of diapsids as not being related to each other. Today, the diapsids are placed into two subclasses: the "advanced" Archosauria (ruling reptiles) and the "primitive" Lepidosauria (scaly reptiles). Supposedly, the birds are closely related to the archosaurs, but this doesn't work because there is greater genetic similarity between birds and mammals, and the birds have only one temporal opening in the skull. Additionally, "primitive reptiles" (diapsids) supposedly evolved into synapsids—some of which evolved backwards, becoming birds and dinosaurs (diapsids). But others supposedly maintained their synapsid anatomical features and evolved into modern mammals.
3. Synapsids (one temporal opening)—"mammal-like reptiles" and mammals. This subclass of reptiles contains the Pelycosauria and Therapsida. The therapsids supposedly replaced the pelycosaurs during the Permian system. The therapsids became extinct in the early Jurassic. Therapsids (e.g., *Dimetrodon*) are supposedly the ancestors of mammals, sharing with them a synapsid skull with one temporal opening.

One secular publication demonstrates the equivocal nature of evolutionary classification.

> Including turtles in Diapsida is *controversial*; however, molecular evidences *suggest* this *interpretation. If* this *interpretation* is correct, the absence of temporal fenestrae in their skulls *must be* a derived characteristic, and resemblance of turtle skulls to ancient anapsids *would be* a result of evolutionary convergence."[17]

Evolutionists liberally use the word "convergence" as an escape in their literature (e.g., it is used about 50 times in Cox and Hautier's 2015 text). It is a problem to phenetic (a branch of numerical taxonomy) classification.

3. The non-Darwinian interpretation: This is not a transition. Kathleen Hunt admits that it was "probably *not* directly ancestral to sharks." Indeed, evolutionists consider this strange shark an oddball. It had a combination of "derived and ancestral" characteristics. Non-evolutionists state *Cladoselache* was a shark with a skeleton made from calcified cartilage as in modern sharks. Under "Cladoselachiformes," Allaby calls *Cladoselache* a fossil shark.[34]

Two evolutionists stated in a 1977 issue of *American Zoologist* only that "a theory of relationships has been proposed":

> In regard to the Paleozoic and Mesozoic elasmobranchs, a theory of relationships has been proposed for six better known genera (*Xenacanthus, Denaea, Cladoselache, Hybodus, Ctenacanthus* and *Paleospinax*) by evaluation of certain characters in the skull, postcranial axial skeleton, fin supports, and fins."[35]

The anatomy of *Cladoselache* is well-preserved down to muscle fibers, indicating *rapid* burial and preservation. To conclude, since *Cladoselache* was a 100 percent shark with jaws and teeth, how could it be a "transition"?

Tristychius (an "ancient" hybodont shark)

> Hunt's Claim: *Tristychius* and similar hybodonts (early Mississippian)—Primitive proto-sharks with broad-based but otherwise shark-like fins.

1. What the fossil evidence actually shows: Fossils show that *Tristychius* was similar to a number of modern sharks (e.g., the cat shark). It was small, fusiform, with a curved, well-developed caudal fin. *Tristychius* also had spikes at the bases of the dorsal fins.
2. The evolutionary interpretation: Evolutionists state that this genus is extinct, its fossils are found in the Carboniferous, and it was related to *Hybodus*.
3. The non-Darwinian interpretation: This is not a transition. *Tristychius* is not even mentioned in works by Colbert, Janvier, Benton, Allaby, or Stahl. It was a fully formed shark, not a transition.

Ctenacanthus (a hybodont shark)

> Hunt's Claim: *Ctenacanthus* & similar ctenacanthids (late Devonian)—Primitive, slow sharks with broad-based shark-like fins & fin spines. Probably ancestral to all modern sharks, skates, and rays. Fragmentary fin spines (Triassic)—from more advanced sharks.

1. What the fossil evidence actually shows: Incomplete fossils have been found. "Their fin spines are *similar to those of modern sharks*: there are two dorsal fin spines, which have a pectinate ornament, and they are deeply inserted into the muscle mass of the body."[36]
2. The evolutionary interpretation: "Probably ancestral to all modern sharks, skates, and rays," according to Kathleen Hunt.
3. The non-Darwinian interpretation: This is not a transition. "That the ancestors of the hybodonts were sharks similar to *Cladoselache* is *suggested but not proved* by inspection of fossils belonging to the genus *Ctenacanthus*."[37]

Other evolutionary comments on this fossil are:

> The skull of most of the ctenacanthids is unknown.[38]
>
> *Ctenacanthus* from the Devonian is poorly known.[39]
>
> The hybodonts survived into the late Cretaceous side-by-side with the modern sharks, the neoselachians.[40]

> Two groups of sharks in the order Galeomorpha, each represented by a small number of genera, *seem to be* connecting links between the hybodonts and the hosts of varied sharks that live in today's oceans.[41]

It should also be noted that *Ctenacanthus* is dated as being older than *Tristychius*, so *Tristychius* cannot be an ancestor of *Ctenacanthus*. Kathleen Hunt can only say "probably" in regard to *Ctenacanthus* and its supposed evolutionary ancestry. Creationists maintain *Ctenacanthus* was 100 percent shark.

Paleospinax (an "early Jurassic" modern shark)

> Hunt's Claim: *Paleospinax* (early Jurassic)—More advanced features such as detached upper jaw, but retains primitive ctenacanthid features such as two dorsal spines, primitive teeth, etc.

1. What the fossil evidence actually shows: This is a "better known genera" of sharks. Some specimens uncovered from Germany and England include well-preserved impressions of vertebrae, jaws, and teeth. It was less than three feet (1 meter) long.
2. The evolutionary interpretation: Evolutionists attempt to place *Paleospinax* as an evolutionary ancestor of modern sharks, but there are uncertainties.
3. The non-Darwinian interpretation: This is not a transition. Although Ms. Hunt claims *Paleospinax* as a transitional form, two evolutionists evidently do not list it so:

> Preliminary cladistic analysis corroborates the hypothesis that the Elasmobranchii and the Holocephali are sister groups, and that the Chondrichthyes are more closely related to the Teleostomi (Acanthodi plus Osteichthyes) than either is to the Placodermi. In regard to the Paleozoic and Mesozoic elasmobranchs, a theory of relationships has been proposed for six better known genera (*Xenacanthus*, *Denaea*, *Cladoselache*, *Hybodus*, *Ctenacanthus* and *Paleospinax*) by evaluation of certain characters in the skull, postcranial axial skeleton, fin supports, and fins.[42]

In regard to the above quote, it should be noted that cladism (a school of thought) considers only evolutionary relatedness. In other words, cladistics presupposes evolution. "More closely related to" is at best a subjective statement. "Hypothesis" and "a theory of relationships has been proposed" are cautious phrases, and *Paleospinax* is simply one of six "better known genera," not a transitional form as proposed by Kathleen Hunt.

Finally, although *Paleospinax* is supposedly a missing link, it is not mentioned in Colbert, Stahl, Benton, or Allaby.

Spathobatis ("proto-ray")

> Hunt's Claim: *Spathobatis* (late Jurassic)—First proto-ray.

1. What the fossil evidence actually shows: Complete rays suddenly appear in the fossil record and are very similar to modern guitarfishes. The following quote from an article in the journal *Palaeontology* shows that the authors feel *Spathobatis* to be a species of ray:

> Sampling of hiatal horizons within the Hauterivian part of the Speeton Clay Formation of northeast England has produced teeth of several species of sharks and rays, four of which are previously unnamed. One species of shark, *Cretorectolobus doylei* sp. nov., and two species of rays, *Spathobatis rugosus* sp. nov. and *Dasyatis speetonensis* sp. nov., are named, whilst the presence of an indeterminate triakid shark is also noted.[43]

2. The evolutionary interpretation: According to Ms. Hunt, *Spathobatis* is the first "proto-ray."
3. The non-Darwinian interpretation: This is not a transition. *Spathobatis* is one of the earliest known fossil rays, but it was very close to modern forms. Underwood et al are quite candid regarding the state of ancient creatures such as *Spathobatis*:

A unique (e.g., teeth unlike those of any other living shark) frilled shark of the genus *Chlamydoselachus* was first described by American chondrichthyan taxonomist S. W. Garman in 1884. *Chlamydoselachus* has an enlarged mouth and eel-like body. Like other hexanchids, it has a single dorsal fin and a long caudal fin. Today, there are five extant (living) species of hexanchiforms, including the frilled and cow sharks.

2. The evolutionary interpretation: Hexanchiforms are considered "an ancient lineage," with one fossil of *Notidanoides muensteri* dating from the late Jurassic ("150 million years ago"). Evolutionists see the hexanchiforms ranging from the early Jurassic ("180 million years ago") to the Tertiary, with some living today.

3. The non-Darwinian interpretation: This is not a transition. Neither Colbert, Allaby, Benton, nor Anderson and Sues refer to *Chlamydoselachus* as a missing link. Stahl has the following to say regarding Ms. Hunt's "transitional form":

   > Since *Chlamydoselachus* also has six pairs [of gill arches] and shows other structural similarities to the Hexanchids, paleontologists suspect some sort of relationship between the two aberrant types of sharks. The specialization that has occurred among the former and changes that have taken place in the jaw structure of the latter seem to prove, however, that the two groups have been evolving separately for a long time.[56]

Summary: Hexanchiforms are 100 percent sharks—not missing links. Virtually all the proposed "transitional sharks" are easily recognized as sharks, showing only that a wide variety of sharks and other cartilaginous fishes existed in the past, just as they do today. There is undoubted variation within the shark kind, as is expected on the basis of creation. None of the supposed transitions suggest that cartilaginous fish evolved from or into any other major group.

Osteichthyes (Bony Fish)

According to evolution, one group of ostracoderms led to the gnathostomes ("jaw mouth"). The bony fish (osteichthyes) supposedly evolved from "primitive" gnathostomes. But the fossil record shows this to be a poorly documented transition: "One of the greatest unsolved events in vertebrate evolution is the origin of Gnathostomata (vertebrates with jaws)."[57]

Contrary to the optimistic list of supposed transitional fish fossils Kathleen Hunt presents, a more somber evaluation is given by two evolutionists:

> Our current knowledge about phylogeny and classification of "fishes" is in a state of flux. Most classification schemes which proposed to organize the vast fish biodiversity have been based on loosely formulated syntheses of many largely disconnected phylogenetic studies among some of its components. An explicit cladistic analysis including representatives of all major taxonomic groups across the diversity of fishes has never been accomplished. As a consequence, phylogenetic relationships among the major groups of fishes are still controversial and unresolved, as are many of the proposed higher-level taxa.[58]

In Figure 2.10 of his text, Benton shows the Placodermi, Acanthodii, Osteichthyes, Chondrichthyes, and Agnatha all appearing separately and suddenly. He states the "dashed lines indicate hypothesized relationships" (i.e., Darwinian evolution).[59] Evolutionists Miller and Harley state, "One interpretation of the phylogeny of the Craniata with emphasis on the fishes. The evolutionary relationships among fishes are unsettled."[60] Colbert in his "Geologic time chart" shows the Agnathans, Placoderms, Acanthodians, Cartilaginous fishes, and Bony fishes appearing suddenly and separately.[61] Dotted lines (which evidently represent the unobserved evolutionary theory) connect the last four groups to the Agnathans.

> The fishes are of ancient ancestry, having descended from an unknown free-swimming protochordate [a tunicate or lancelet] ancestor about 500 million years ago.[62] [Note that this is a faith claim, *not a fact of science*.]
>
> [Fishes] appear in the fossil record in the late Silurian period with fully formed jaws, and no forms intermediate between Agnathans [jawless] and gnathostomes [jawed] are known.[63]
>
> The higher fishes, when they appear in the Devonian period, have already acquired the characteristics that identify them as belonging to one or another of the major assemblages of bony or cartilaginous forms.[64] [Since they first appear as distinctly separate groups, is it fair to ask whether they "acquired the characteristics that identify them" by creation rather than by descent from unknown and undiscovered ancestors?]
>
> Both these groups—bony and cartilaginous—appeared in the late Silurian period, and it is possible that they may have originated at some earlier time, although there is no fossil evidence to prove this.[66]

Colbert (above) is indeed correct, not only is it "possible that they may have originated at some earlier time"—but 100 percent fish fossils have now been discovered in China.[67] These fully-finned, jawed fish (*Guiyu oneiros*) have been unearthed in rock that has been evolutionarily dated three million years earlier than the late Silurian fossils ("416 million years ago") to which Colbert refers. A University of Chicago evolutionist responds to this recent discovery:

> By pushing a whole series of branching points in gnathostome [jawed vertebrates] evolution out of the Devonian and into the Silurian, the discovery of *Guiyu* also signals that a significant part of early vertebrate evolution is unknown.[68]

The reader should note that this "earliest" fish fossil (*Guiyu oneiros*) is 100 percent fish and not a transitional form. Evolution-based museum dioramas incorrectly displaying fishless Silurian seas must now be populated with fish.

Transition from primitive bony fish to holostean fish

Most fish living today have rays in their fins and some have bone. Kathleen Hunt calls the bony-finned fish "primitive" and claims a few forms to support her belief that some of these evolved into "advanced" ray-finned (i.e., holostean) fish. Ms. Hunt states, "Paleoniscoids...*probably* gave rise both to modern ray-finned fish...and also to the lobe-finned fish" (emphasis added). Statements by other evolutionists indicate the tenuous nature of these claims:

> There seems to be as little chance of tracing the origin of the paleoniscoids as of any other kind of fishes, but investigators *speculate* on the matter, nevertheless. They debate not only the relationship of the first ray-fins to the acanthodians but also the type of habit in which the line developed.[69]

Acanthodians are "primitive" fossil fish with a true bony skeleton, large eyes, prominent spines, and a very imprecise evolutionary past.

> The clade Acanthodii no longer exists....So there has either been massive convergence between different acanthodians and other taxa, or acanthodians are not a clade.[70]

Cheirolepis (an early actinopterygian, the large group including most familiar ray-finned fish)

> Hunt's Claim: In her separate Transitional Fossils FAQ, Ms. Hunt claims that the transition from "primitive" bony fish to holostean fish involved Palaeoniscoids (e.g., *Cheirolepis*); living chondrosteans such as *Polypterus* and *Calamoichthys*; and also the living acipenseroid chondrosteans such as sturgeons and paddlefishes, followed by primitive holosteans such as *Semionotus*.

1. What the fossil evidence actually shows: Discovered in the Upper Devonian in eastern Canada, *Cheirolepis* was a fast-swimming predatory, ray-finned (bony) fish (9.5 inches long) with tiny scales. Supposedly acanthodian-like, their fossils are known from all parts of the world.
2. The evolutionary interpretation: Evolutionists assign *Cheirolepis* to mid-Devonian. Ms. Hunt calls this a "transition from primitive bony fish to holostean fish." Anderson and Sues consider it "more advanced," saying *Cheirolepis* "is often considered as the most basal actinopterygian apart from *Dialipina*."[71]
3. The non-Darwinian interpretation: This is not a transition. Barbara Stahl mentions *Cheirolepis*, but not as a transitional form.[72] The language of Colbert is cautious: "*We might say* that *Cheirolepis*, an early actinopterygian...had various characters in common with *Dipterus*, a generalized lungfish, and *Osteolepis*, a generalized ancestral crossopterygian."[73] Allaby states *Cheirolepis trailli* is "an early representative of the primitive bony fish, the palaeoniscids known from the Middle Devonian."[74] He did not call it a transitional form.

Benton discusses *Cheirolepis* at length but does not call this creature a transition.

> Devonian actinopterygians such as *Cheirolepis* are known from all parts of the world, but only some 15 genera have been found so far. The actinopterygians radiated dramatically in the Carboniferous and later, and they are the dominant fishes in the seas today.[75]

Polypterus—an African bichir (Brachiopterygii)

> Hunt's Claim: *Polypterus* is part of the transition from bony fish to holostean fish (see above claim).

1. What the fossil evidence actually shows: There are 11 living species of this freshwater fish of the family Polypteridae and a number of fossil species.
2. The evolutionary interpretation: Kathleen Hunt states this fish is a "transition from primitive bony fish to holostean fish."
3. The non-Darwinian interpretation: This is not a transition. Six evolutionary zoologists stated only that "*Polypterus bichir*, of equatorial West Africa...is a nocturnal predator."[76] But the authors did not mention *Polypterus* as being a missing link. Benton stated, "The bichirs, Polypteridae, are heavily armored fishes that live in the

streams and lakes of tropical Africa, and famous as so-called 'living fossils' that evolved slowly and at low diversity."[77]

Barbara Stahl refers to *Polypterus* as a degenerate remnant "of the paleoniscoid group."[78] Degenerate does not mean upward and onward evolution. Michael Allaby says that *Polypterus* is "curious" and that it has "specialized features,"[79] but he does not label it as a missing link. He also states, "The systematic position [systematics attempts to explain biological diversity in an evolutionary context] of the bichirs is not yet certain."[80]

Colbert too does not claim *Polypterus* as a missing link and states only that it and *Calamoichthys* of Africa were "supposed to be of crossopterygian relationships."[81] Under the title 'Transition from Fin to Limb," Anderson and Sues display an unconvincing phylogenetic diagram of the Sarcopterygii that includes *Polypterus*. The theme of degeneration is also mentioned here when the authors state that *Polypterus* has *lost* endochondral elements (radials).[82]

Calamoichthys (reedfish)

Hunt's Claim: *Calamoichthys* is part of the transition from bony fish to holostean fish (see above claim).

1. What the fossil evidence actually shows: These fish belong to the Brachiopterygii (class Osteichthyes).
2. The evolutionary interpretation: Allaby states "the systematic position of the bichirs [Brachiopterygii] is not yet certain."[83] Long states, "There are two living representatives of this early actinopterygian radiation, the freshwater African reedfishes, *Polypterus* and *Calamoichthys*."[84]
3. The non-Darwinian interpretation: This is not a transition. Benton, Clack, and Stahl do not mention *Calamoichthys*. Long does not call this fish a transitional form. Colbert mentions this genus along with *Polypterus* as both belonging to the order Polypteriformes. Colbert also states that Polypteriformes was "for many years supposed to be of crossopterygian relationships."[85]

Twenty-first century evolutionary literature does not support Hunt's claim.

Acipenseroid chondrosteans (sturgeons and paddlefishes)

Hunt's Claim: Acipenseroid chondrosteans are part of the transition from bony fish to holostean fish (see above claim).

1. What the fossil evidence actually shows: Sturgeons and paddlefishes are alive today.
2. The evolutionary interpretation: Colbert states, "Then, at this stage of earth history, there were the early sharks exploring the possibilities of life in the sea, some of them sharing the environment with the giant arthrodires. Finally, there were the primitive bony fishes, including the first actinopterygians. These were for the most part freshwater fishes at first."[86]
3. The non-Darwinian interpretation: This is not a transition. Benton says, "The phylogeny [evolution] of the Devonian to Triassic actinopterygians has proved hard to establish."[87] Colbert shows the Chondrosteans (e.g., sturgeon) appearing suddenly in the early Triassic.[88]

Semionotus (an early Mesozoic holostean, class Osteichthyes)

Hunt's Claim: *Semionotus* is part of the transition from bony fish to holostean fish (see above claim).

1. What the fossil evidence actually shows: There are both living and fossil representatives of the order Semionotiformes. They "occur in great diversity in some areas, such as the Newark Group (Late Triassic and Early Jurassic) lakes of the eastern seaboard of North America, where they appear to have formed species flocks, in which numerous species lived together."[89]

2. The evolutionary interpretation: Barbara Stahl considers *Semionotus* one of the transitions from primitive bony fish to holostean fish.

3. The non-Darwinian interpretation: This is not a transition. Allaby states the Semionotiformes are "an order of bony fish that has fossil and living representatives" but does not call them a transitional form.[90] Stahl is candid regarding semionotid origin:

> When structural similarity is not close, paleontologists cannot carry their speculations very far. For this reason, the origin of the semionotids can only be debated, and the source of the pholidophorids, holostean forerunners of the teleosts, can hardly be supposed until subholosteans more like them are found.[91]

Then, in 1982, in an article titled "Correlation of the early Mesozoic Newark Supergroup by Vertebrates" in the *American Journal of Science*, the authors—under the subheading "The semionotid fish zones"—state:

> Semionotid fishes are the most common vertebrates in the Newark Early Jurassic, and the restriction of some forms to certain fish-bearing beds in several basins makes them useful for correlation. Unfortunately, taxonomic confusion at both specific and generic levels prevents the use of formal systematic nomenclature at this time.[92]

At no point in this *American Journal of Science* article do the authors address *Semionotus* as a transitional form.

Speculating only on conflicting claims supported by some parts of some fossils and refuted by others, evolutionists have not been able to agree on answers to "little questions" about how some fish may have evolved into other fish. Will they do better with a "big question," such as how some fish became a land animal?

Transition from primitive bony fish to amphibians

When the fossil record fails to document the transition from water to land, secular textbook writers are forced to concoct "just-so" stories. Hence, we see in the following paragraph the author advocating that the fish somehow anticipates it will need limbs millions of years in the future:

> Sometimes, by chance, an organ that works well in one function turns out to work well in another function after relatively little adjustment. Fins in both fish groups evolved for swimming. In some lobe-finned species, they probably came to be used for scuttling around near the seashore or on the bottom of rivers or lakes. From this point, only a small change was required for the fish to walk on land. Whatever the details involved, it is a reasonable inference that the lobe-finned skeleton was, unlike a ray-fin, preadapted to evolve into a tetrapod limb. The term preadaption is applied when a large change in function is accomplished with little change of structure.[93]
>
> No one can be certain which group or groups of fishes was the first to make the transition to land, or what their evolutionary pathways may have been...the transition from water to land occurred so long ago, and various family trees suggested by the fossil record are so tangled that scientists acknowledge they may never be able to sort them out definitively.[94]

Osteolepis (a Devonian sarcopterygian)

> Hunt's Claim: *Osteolepis* (mid-Devonian)—One of the earliest crossopterygian lobe-finned fishes, still sharing some characters with the lungfish (the other lobe-finned fishes). Had paired fins with a leg-like arrangement of major limb bones, capable of flexing at the "elbow", and had an early-amphibian-like skull and teeth.

1. What the fossil evidence actually shows: *Osteolepis* possessed an intracranial joint and heterocercal tail (vertebral column extends into the tail) with a larger lower lobe and two posteriorly placed rounded dorsal fins. Complete fossils of this creature have been found at a famous Scottish site showing good quality rhombic scales.

2. The evolutionary interpretation: Evolutionists suggest Osteolepiform fish are the ancestors of the Tetrapoda (four-limbed animals) because of their paired lobed fins (see Long, 2011). *Osteolepis* is not mentioned in Allaby or Hickman et al.

3. The non-Darwinian interpretation: This is not a transition. Colbert subjectively states that the osteolepiforms had "a skull pattern remarkably prophetic of the skull pattern seen in the early amphibians."[95] But Benton does not refer to this osteolepiform as a transition, calling it a lobefin and fish.[96] He went on to state:

   > The expectation is, of course, that [coelacanths, lungfishes, and tetrapods] group as Sarcopterygii to the exclusion of Actinopterygii, but relationships between coelacanths, lungfishes, and tetrapods are debated. It has been surprisingly difficult to resolve the three-clade problem within Sarcopterygii."[97]

Clack calls *Osteolepis* "one of the most primitive osteolepidids from the Middle Devonian" and an "early lobe-fin," but does not state it was a transition.[98] Stahl documents the complicated and difficult debate between evolutionists (e.g., Romer, K.S. Thomson, and Jarvik) regarding where to place *Osteolepis* and why.[99]

Sarcopterygians (Fleshy-Finned Fish)

Among the fleshy-finned sarcopterygians are lobe-finned fish and lungfish. A majority of evolutionists see the fossil lobe-finned crossopterygians (coelacanths and fossil relatives) as possible ancestors of tetrapods, but none are known to be capable of either walking or breathing out of water. The *Dipnoi* (lungfish) can use their swim bladders to absorb oxygen from air and have been separated from lobe-finned fish.

> The Rhipidistia have been traditionally thought to be the tetrapod ancestors, on the basis of overall similarity in the skull roof, appendages, and approximate stratigraphic position. In particular, the presence of paired internal nostrils (choanae) has been cited as a linking character. Rosen, Forey, Gardiner, and Patterson (1981) claimed that the interpretation of this character is incorrect and that the rhipidistian *Eusthenopteron* lacks choanae. By contrast, a restudy of a Devonian lungfish from Australia suggests the presence of choanae. Thus, the restudy of characters placed the Dipnoi (lungfish) as the sister group of the tetrapods and completely changed our conception of vertebrate [evolution].[100]
>
> The question of where tetrapods evolved is even more difficult to answer than that of when.[101]

There have been fossil discoveries such as *Psarolepis, Ligulalepis,* and *Dialipina* that show fascinating combinations of osteichthyan and non-osteichthyan gnathostome characters. For the evolutionist this is confusing, to say the least, with *Psarolepis* having traits that are found in lobe-fins, chondrichthyans, ray-finned fishes, and Placoderms.[102]

An even earlier specimen has been discovered that places a fully boned, jawed fish "three million years" back into the Silurian![103] It is called *Guiyu oneiros* and is unquestionably sarcopterygian and 100 percent fish, having some unique characters. *Guiyu* had eyes, teeth, gills, scales, and fins.

In addition, one should keep in mind a discovery of a "primitive" 415 million-year-old bony fish (*Janusiscus piscus*) with traits of sharks that complicates the evolutionary story of fish ancestry and "may rewrite [the] fish family tree."

> Although the fossil had previously been classified as a bony fish based on its external features, such as the shape of the skull roof and the enamel on the scales, the CT scan revealed a surprising mosaic of features from both cartilaginous and bony fish. For example, the fish's skull was made of large, bony plates similar to today's bony fish, but the traces of the nerves and blood vessels around the brain more closely resembled those of cartilaginous fish.[104]

Order Rhipidistia (e.g., Eusthenopteron)

> Hunt's Claim: *Eusthenopteron*, *Sterropterygion* (mid-late Devonian)—Early rhipidistian lobe-finned fish roughly intermediate between early crossopterygian fish and the earliest amphibians. *Eusthenopteron* is best known, from an unusually complete fossil first found in 1881. Skull very amphibian-like. Strong amphibian-like backbone. Fins very like early amphibian feet in the overall layout of the major bones, muscle attachments, and bone processes, with tetrapod-like tetrahedral humerus, and tetrapod-like elbow and knee joints. But there are no perceptible "toes", just a set of identical fin rays. Body & skull proportions rather fishlike.

1. What the fossil evidence actually shows: The rhipidistians (e.g., *Eusthenopteron*) are a group of bony fish with patterns of bones in the fins. Fossils have been found of a creature with fairly well-defined skull bones, teeth, and vertebral column.
2. The evolutionary interpretation: The remains of ancient lobe-finned fishes placed in the order Rhipidistia are seen by a majority of evolutionists today as the "ideal" amphibian ancestors. They are found in the late Devonian and have been classified as an intermediate between terrestrial vertebrates and fish. Coelacanths are considered in or close relatives of the order Rhipidistia. *Sterropterygion* is not mentioned in Allaby, Benton, Clack, or Hickman, et al.

3. The non-Darwinian interpretation: *Eusthenopteron* and *Sterropterygion* are not transitions. Science has shown the bony portions of fins of the rhipidistians are loosely embedded in muscle and not attached to the vertebrae. Yet the oldest amphibian (according to evolutionists) has strong shoulders and a pelvis, designed to support the weight of the amphibian while on land. Every fish, living or fossil, even those with unusual characteristics, is fully fish, and every amphibian, fossil or living, is fully amphibian. Not surprisingly, it has been found that the genomes of ray-finned fishes "might be considerably different" from their supposed "sister group," the land vertebrates—but this is attributed to genetic events supposedly occurring "in their evolutionary past."[105]

> None of the known fishes is thought to be directly ancestral to the earliest land vertebrates. Most of them lived after the first amphibians appeared, and those that came before show no evidence of developing the stout limbs and ribs that characterized the primitive tetrapods.[106] [Anti-creationists maintain this is a dated book and that "there have been quite a few discoveries since" Stahl wrote this—but recent discoveries have failed to confirm vertebrate evolution.]
>
> [Rhipidistia] are considered *by some* to be ancestral to the tetrapod, terrestrial vertebrates (i.e. Amphibia).[107]
>
> *For the moment*, it is hard to see that *opinion* will be *swayed away* from the *idea* that tetrapods are the descendants of a tetrapodomorph lineage including fish such as *Tiktaalik, Panderichthys, Eusthenopteron,* and *Osteolepis.*[108]
>
> As far as the fin-limb transition is concerned, Rosen *et al.* argued that the dichotomous pattern of the paired fin skeleton in *Eusthenopteron* is merely the structure of the metapterygium of other gnathostome fishes, and thus is general for sarcopterygians. In consequence, it could not be regarded as closer to the structure of the tetrapod limb in this respect.[109]

In the 21st century, the Rhipidistia has had to be redefined:

> These efforts have culminated in complete reworking of the relationships of lobe-fins, such that terms like "crossopterygian" and "rhipidistian" are no longer used with their original meanings. As they had come to be defined over the years, they embraced what cladists call *paraphyletic groups* held together only by what they lacked, and part of this problem was caused by the exclusion of the lungfishes. The lungfishes were now returned to the fold, and in most recent cladistic analyses, it is the lungfishes rather than the coelacanths that emerge as the tetrapods' closest living relatives.... Part of the problem stems from the fact that all three groups—lungfish, coelacanths, and tetrapods—have a long evolutionary history separate from one another, which means that they are all rather different from their early ancestors.[110]

That is, *if they had early ancestors*. Non-evolutionists maintain these three groups have obviously been created as separate groups and had no evolutionary history.

> The other group of crossopterygians, the coelacanths, was far removed from the line of evolution that led toward the early land vertebrates.[111]
>
> Hence, when the modern coelacanth *Latimeria* was found, as a living crossopterygian, it was hailed as the surviving embodiment of a tetrapod ancestor. The problems that its anatomy eventually posed for this theory have been outlined above.[112]
>
> [T]here are at least four major structural differences between lobefinned fish and the early amphibians: the attachment of fore and hind limbs, the orientation of the limbs, and the formation of polydactyl [having multiple fingers or toes] limbs. There are no fossils between the fish and amphibians showing creatures having only some of these features, nor of any having an intermediate stage for any of these structures, e.g. in the formation of a polydactyl limb.[113]
>
> Unfortunately, the relationships between lungfishes, coelacanths, and tetrapods are not so easily established as they first appeared, and the jury is still out 30 years later.[114]

There are significant differences regarding early amphibians and the lobe-finned fish. Specifically, the orientation of the fore and hind limbs, their attachments, and the formation of polydactyl limbs. When it comes to the relationships of the basal tetrapods, Benton states, "The assignment of the major extinct clades of basal tetrapods—the temnospondyls, lepospondyls, embolomeres, seymouriamorphs, and diadectomorphs—has been controversial,"[115] and Hickman et al agree: "Evolutionary relationships of early tetrapod groups remain controversial."[116]

Eusthenopteron (a Devonian crossopterygian fish, once presumed ancestral to Icthyostega)

> Hunt's Claim: *Eusthenopteron* (and other rhipidistian crossopterygian fish)—intermediate between early crossopterygian fish and the earliest amphibians. Skull very amphibian-like. Strong amphibian-like backbone. Fins very like early amphibian feet.

1. What the fossil evidence actually shows: Classified as Osteichthyes, subclass Crossopterygii, all its fins were true fins, but it also had five radial bones (*found only in fish fins*) connected to the pelvis. It had a symmetrical tail with a vertebral column that extended straight back. *Eusthenopteron* evidently had bones connecting the rear fins to the backbone.
2. The evolutionary interpretation: This was the "gold standard" of fish with supposed legs, according to Darwinists. The creature is from the late Devonian and is, according to Colbert, "of particular importance." He states, "The crossopterygians are to us perhaps the most important of fishes; they were our far-distant but direct forebears."[117] Some evolutionists feel this creature may be the true "missing link" instead of the coelacanth. Allaby states a number of advanced characters suggests that "it was closely linked with the evolution of the Amphibia. It was related to the lungfish and coelacanths."[118]
3. The non-Darwinian interpretation: This is not a transition. Placing *Ichthyostega* (one of the "earliest amphibians") alongside *Eusthenopteron* shows them to be eons apart anatomically. The portion of *Eusthenopteron* considered by evolutionists to be like that of a land animal was *not* the body (e.g., fins) but the head region—which has features that are frog-like, according to Jarvik.[119] "Our fingers and toes really did evolve from the fins of ancient fish…but they do not appear to have any bones that could have gone on to produce digits."[120]

Eusthenopteron was a fish, having no legs or feet. The fins were true fins with a distinctively fish-like arrangement. Non-Darwinian zoologists ask how the gills of *Eusthenopteron* changed into the lungs of *Ichthyostega*. By a series of beneficial mutations, a pelvis must evolve in *Ichthyostega* from where there was no pelvis in *Eusthenopteron*. Clack notes *Eusthenopteron* "was much more like that of a modern pike (*Esox*), *a fully aquatic ambush predator*."[121] Most paleontologists agree *Eusthenopteron* was not on the evolutionary path to tetrapods. Benton states, "*Eusthenopteron* could not have walked properly on land on its fins."[122] Evolutionists state the bony elements of *Eusthenopteron* "foreshadowed" bones of tetrapod limbs, but in the end

> Understanding the evolutionary transformation of fish fins into tetrapod limbs is a fundamental problem in biology. The search for antecedents of tetrapod digits in fish has remained controversial because the distal skeletons of limbs and fins differ structurally, developmentally, and histologically. Moreover, comparisons of fins with limbs have been limited by a relative paucity of data on the cellular and molecular processes underlying the development of the fin skeleton.[123]

There is a total lack of fossil evidence documenting the supposed appearance of a pectoral girdle from the tristichopterid fish to a basal tetrapod. Benton stated regarding *Eusthenopteron* that "new bones appeared."[124] Appealing to the appearance of morphological structures is not science—it's engaging in magic.

Another example is found in Clack:

> In effect, the head is no longer supported by the shoulder girdle, so the vertebral column and its muscles must do the job instead. Thus, tetrapods have necks.[125]

In the next paragraph she states, "The back of the tetrapod head *became gradually adapted* to provide anchorage and space for these muscles to attach."[126] These are not scientific descriptions, of course. This applies to the pelvic girdle as well: "The pelvic girdle was also much modified."[127]

All three subdivisions of the bony fishes appear in the fossil record at approximately the same time....How did they originate? What allowed them to diverge so widely?...And why is there no trace of earlier, intermediate forms?[128]

Most paleontologists agree *Eusthenopteron* was a fish.

Tiktaalik

Claim: Although not listed in Kathleen Hunt's FAQ because it wasn't discovered yet, *Tiktaalik* has been widely touted as an intermediate form between fish and amphibians.

1. What the fossil evidence actually shows: Three fossilized *Tiktaalik roseae* skeletons were found in rock (formerly river sediments) on Ellesmere Island, northern Canada. The discovery was made by a paleontologist who noticed one of the skulls protruding from a cliff and was published in the April 6, 2006, issue of *Nature*.[129]
2. The evolutionary interpretation: Alleging it to be a "transitional form" between fish living about 385 million years ago (MYA) and early tetrapods living approximately 365 MYA, Neil Shubin, one of the discoverers, proclaimed *Tiktaalik* a "fishapod" and made it the centerpiece for his book *Your Inner Fish*.[130] A new study of *Tiktaalik roseae* highlights an intermediate step between the condition in fish-like *Eusthenopteron* and that in early limbed forms like *Acanthostega*.[131]
3. The non-Darwinian interpretation: This is not a transition. Previous candidates that were touted and taught without question for decades as "proof" for fish-to-amphibian evolution have now been discarded. There is no doubt there have been "late Devonian" lobe-finned fish and amphibious tetrapods, but are there links that would connect these two very different groups?

The secular media addresses *Tiktaalik* as evidence against creation science. *The New York Times* describes *Tiktaalik* "as a powerful rebuttal to religious creationists, who hold a literal biblical view on the origins and development of life."[132] However, there is no reason to assume that *Tiktaalik* was anything other than an exclusively aquatic animal. Just because the evolutionists call the appendages "legs" doesn't necessarily make them legs. Suspicions continue, with evolutionists giving grossly conflicting dates for this creature. (The late Devonian is considered between 385-359 MYA.) In December 2008, *Nature* published a letter by three European evolutionists stating that *Tiktaalik* is not a missing link after all and that the situation is a lot more complex (i.e., *Panderichthys'* fin may be anatomically closer to tetrapods than *Tiktaalik*).[133]

Another problem is the discovery in Australia of a supposed 380-million-year-old "ultimate 'Mother' of all tetrapods" called *Gogonasus*.[134] One of the challenges is that Darwinists want *Gogonasus* to represent an earlier candidate for a tetrapod transition, yet it shares some similarities to the later *Tiktaalik*:

> The conspicuously large spiracular opening (Fig. 1a-c) is proportionally similar to those recently reconstructed for *Panderichthys* and *Tiktaalik*. The pectoral fin endoskeleton of *Gogonasus* is described here for the first time (Fig. 2), the new specimen being the only known Devonian fish that shows a complete acid-prepared pectoral limb. There are some surprising similarities to the recently described pectoral fin in the advanced elpistostegalian *Tiktaalik*. As such features could indicate homoplasy between *Gogonasus* and early tetrapods, we present a revised character analysis to determine whether the new anatomical information supports a more crownward position for *Gogonasus* in the stem-tetrapod phylogeny.[135]

Note that the presumptive idea of homoplasy (convergent evolution) is used to explain why this early fish would have similar structures to a later one.

In early 2010, clear, vertebrate trackways said to be "18 million years" older than *Tiktaalik*, showing digits and alternating steps, were announced in *Nature*. The article states, "They force a *radical reassessment* of the timing, ecology and environmental setting of the fish–tetrapod transition, as well as the *completeness of the body fossil record*."[136] Two evolutionists commented, "The fish-tetrapod transition was thus seemingly quite well documented....Now, however, Niedzwiedzki et al *lob a grenade into that picture*."[137]

With no apologies to *The New York Times*, creationists state that the fossil record is a powerful rebuttal to secular evolutionists who hold a literal Darwinian view on the origins and development of life.

1. What the fossil evidence actually shows: In 1897, Swedish scientists fortuitously discovered this fossil in East Greenland. Of all the early tetrapods, the fossil of *Icthyostega* is the most thoroughly studied, thanks to Erik Jarvik, who made it his life's work. The solidly constructed skull was about 6 inches (15 cm) in length.
2. The evolutionary interpretation: *Icthyostega*, a contemporary of *Acanthostega*, had fully formed tetrapod limbs and must have been able to walk on land. The hindlimb bore seven toes (the number of front limb digits is unknown).
3. The non-Darwinian interpretation: This is not a transition.

> The interrelationships of early tetrapods constitute a problem that for the time being is unsatisfactorily resolved. There seems to be little obvious connection between the tetrapods of the Late Devonian [e.g., *Icthyostega*] and those of the early Carboniferous.[156]
>
> The relationships of the Late Devonian tetrapods (see cladogram), and their closest fish relatives, are controversial, not least because many of the specimens are incomplete and are currently under study.[157]

Kathleen Hunt's fellow evolutionists disagree with her pronouncement of *Ichthyostega*'s transitional status. Indeed, transitional forms between the Rhipidistian fish and this "earliest amphibian"—*Icthyostega*—are lacking. For example, missing from evolutionary literature is a cogent description and discussion regarding evolution of the pelvic region. (This subject is not discussed in Colbert's text.) In fact:

> The "four-legged fish" *Ichthyostega* is not the "missing link" between marine and land animals, but rather one of several short-lived "experiments" according to evolutionists from Cambridge & Uppsala Universities....It isn't easy to interpret the fossil of *Ichthyostega*. Even though almost the whole skeleton is represented, there is no single fossil that shows the whole animal. Instead it is necessary to assemble a puzzle from information found in several different fossils.[158]

Clack states:

> As new information comes to light, many of these ideas will be tested and may have to be rethought. Examples of this have already begun to raise questions about when and how terrestriality was acquired, such as the existence of the trackways of tetradomorphs from the early Middle Devonian, and the primitive, fishlike characteristics of the humerus of *Icthyostega*. Clearly the story is more complicated than the current range of fossils indicates.[159]

After a 10-page description of *Icthyostega*, Clack is cautious, saying:

> *Icthyostega* is a curious mixture of features, some of them primitive but some of them specialized and unique.[160]

In 2003, Clack and fellow authors reported in *Nature* a distinct specialization of *Icthyostega* that is not transitional.[161] The anatomy of the unique ear region is a challenge to interpret and it would seem to disqualify this animal as an evolutionary link.

Pholidogaster and Pteroplax (Labyrinthodonts)

Although there are numerous websites that have copied and pasted Kathleen Hunt's comments regarding these two creatures, they are not mentioned in Stahl, Anderson and Sues, or Benton. Allaby states, "The modern tendency is to regard the labyrinthodonts as an informal assemblage of primitive amphibians, rather than as a formal taxonomic group."[162] Even Ms. Hunt's claim for these two "transitionals" actually points out how fully amphibian they really are.

> Hunt's Claim: Labyrinthodonts (e.g., *Pholidogaster*, *Pteroplax*)—still have some icthyostegid features, but have lost many of the fish features (e.g., the fin rays are gone, vertebrae are stronger and interlocking, the nasal passage for air intake is well defined).

Note the verbs in Ms. Hunt's claim: "have lost," "are gone," "are stronger." Such verbs conjure images useful in storytelling, salesmanship, and propaganda, but the hard facts observable to paleontologists are the amphibian features that labyrinthodont amphibians actually have. It is only a faith based on *features not seen*—but only claimed to be "lost" or "gone"—that supports these evolutionary conclusions.

Early Reptiles

> At present, the most likely sister group to the amniotes [containing reptiles] is the diadectomorphs such as the hulking and pig-sized *Diadectes*....The cotylosaurs, meaning literally "stem reptiles," were envisioned to be the basal group of amniotes from which all later groups issue.[163]

In her Transitional Fossils FAQ, Kathleen Hunt includes an "early" reptile group in her list of transitions from reptiles to mammals.

Dimetrodon (Permian Pelycosaurian reptile)

> Hunt's Claim: Therapsids (e.g. *Dimetrodon*)—the numerous therapsid fossils show gradual transitions from reptilian features to mammalian features.

1. What the fossil evidence actually shows: Fairly complete skeletons have been found ranging from 6 to 11 feet long.
2. The evolutionary interpretation: Pelycosaurs are a reptile order of synapsids with vague relationships ("Upper Carboniferous to Lower Permian") that is supposedly ancestral to mammals. *Dimetrodons* theoretically preceded dinosaurs by some 40 million years. Evolutionists feel this animal was replaced by the Therapsida due to teeth and skull morphology.
3. The non-Darwinian interpretation: This is not a transition. One should be aware that there are substantial differences between early pelycosaurs and later therapsids, and that the field of paleontology does not document transitions from one to the other. In regard to the famous "sail lizard," it suddenly appears in the fossil record complete and fully formed as a *Dimetrodon*. It is interesting that evolutionists see this animal as a possible link to mammals, skipping over the entire supposed "dinosaur age" and links to other reptiles.

Mammals

In March 2007, *Nature* magazine published an enormous study regarding mammal origins and radiation.[164] The results were sobering for the evolutionist. What they had taught in schools, museums, and the media for decades in regard to supposed mammal evolution after the great dinosaur extinction "65 million years ago"...never happened. Ross MacPhee of the American Museum of Natural History, co-author of the study, was very surprised. There was no burst of evolution among horses, people, rodents, elephants, or cats after the dinosaur's demise. This is a challenge to the science of paleontology. Not only are the critical transitions between major groups missing, but now secular scientists are back to their bleak search for fossils that reveal information supporting an evolutionary history for mammals.

Editors Anderson and Sues have noted that "the phylogenetic tree of Mesozoic mammals, as supported by the current evidence, is a vast evolutionary bush."[165] Figure 9.1 of their text shows a significant shift from traditional evolutionary interpretation of mammal origins. The non-Darwinian interpretation predicts bushes revealing a mosaic patchwork, each with limited variation. In other words, *a bush does not support the traditional idea of macroevolution, but does support the creation model.*

Six evolutionists described what they think the earliest mammals were like:

> The earliest mammals were *almost* certainly endothermic, although their body temperature *would have been* rather lower than modern placental mammals. Hair was essential for insulation, and the presence of hair implies that sebaceous and sweat glands must also have evolved at this time to condition hair and to facilitate thermoregulation. *The fossil record is silent* on the appearance of mammary glands, but they *must have* evolved before the end of the Triassic. The young of early mammals *probably* hatched from eggs in a very immature condition.[166]
>
> The first mammals appeared in the Late Triassic, but the first fossils are incomplete. *Adelobasileus* and *Sinoconodon* appear to be the most basal mammals.[167]
>
> The presence of hair in mammals may seem a trivial matter, but the exact timing of the origin of hair in mammals is unknown because the fossil record of the evolution of hair in mammals is exceedingly sparse. The developmental origin of hair is equally mysterious; to date, no well-tested (or testable) scenario has been put forth to explain what transformations may account for the evolution of hair.[168]

To the continuing consternation of evolutionists, both hair and feathers develop from complex follicles deep in the dermal layer of skin, quite unlike reptilian scales—their presumed precursors. Milk from mammary glands, the source of the name "mammals," is even more diagnostic of the group than hair. Although they don't often say it out loud, evolutionists believe breasts evolved from sweat glands (missing in reptiles) and milk from sweat. But they don't usually describe what happened to mammal babies during the long, slow, step-by-step transition from a thin, watery, salty solution of urea and various toxins to a thick, nutritious liquid rich in protein, sugar, and fat, and even disease-fighting antibodies.

Morganucodon

> Hunt's Claim in her Transitional Fossils FAQ: Morganucodonts (e.g. *Morganucodon*)—early mammals. Double jaw joint, but now the mammalian joint is dominant (the reptilian joint bones are beginning to move inward; in modern mammals these are the bones of the middle ear).

1. What the fossil evidence actually shows: Morganucodonts were mouse- or shrew-like creatures. Like most modern mammals, this animal possibly had two sets of teeth in its lifetime. The mandible had a trough for post-dentary bones, but the skeleton is poorly known.
2. The evolutionary interpretation: *Morganucodon* (and *Kuehneotherium*) represents the most definitive transitional forms between reptiles and mammals, possessing the mammal-type jaw-joint side by side with the reptile-type jaw-joint. *Morganucodon* had a single ear bone and a full complement of reptilian bones in the lower jaw.

3. The non-Darwinian interpretation: This is not a transition. Evolutionists maintain all mammals (including the blue whale and bat) came from the mouse-like *Morganucodon*. Once again, however, molecular analysis casts much doubt regarding the supposed evolution of mammals. In 2002, evolutionist Mark Springer of the University of California at Riverside stated to *National Geographic News*:

> Our molecular results suggest that living placental mammals have a common ancestor about 105 million years ago....[Other] molecular data also suggest a much earlier split between living placentals and living marsupials—about 175 million years ago.[169]

Creationists ask which molecular evidence will evolutionists use and why—keeping in mind the massive 70-million-year difference. Indeed, five years later it was decided that a figure of "65 million years" is closer to placental mammal origin!

> The controversy is debated not just among paleontologists who study the fossil record but among molecular systematists who study DNA in living mammals. Yet the DNA studies do not agree on the timing or place of placental origin, with hypotheses ranging between 140 and 80 million years ago, sometimes in the Northern Hemisphere and sometimes in the Southern. The most recent molecular study, published in late March in *Nature*, supported the emergence of the major groups of modern placentals 100 million years ago. "Our research gives credence and weight to the traditional paleontological view of placental mammals appearing 65 million years ago when the dinosaurs died off," said Dr. Wible. "When dinosaurs became extinct, ecological niches emerged that gave modern placental mammals opportunities to thrive and diversify."[170]

The *National Geographic News* article that quoted Mark Springer also stated "that indeed there was great mammalian diversity 125 million years ago." Creation scientists agree on diversity (but not "125 million years ago"). Mammals never evolved from a supposed common ancestor and were instead created as a varied group.

> Two new Jurassic fossils yield conflicting reconstructions of the mammalian tree. These divergent genealogies have profoundly different implications for the origin and early diversification of mammals.[171]

See also Benton, Box 10.3 "Relationships of the Mesozoic mammals."[172]

A fossil has yet to be unearthed that represents an intermediate stage, such as one possessing two bones in the ear and three bones in the jaw. Since reptiles and birds with one middle ear bone hear every bit as well as mammals, the selective advantage of the extra bones has never been convincingly articulated, nor the step-by-step development of design features required for the three bones to work together. One revealing study stated:

> The most striking characteristic of the assessory jaw bones of *Morganucodon* is their cynodont character. Compared with such a typical advanced cynodont as *Cynognathus,* the accessory bones present show no reduction, either in size or complexity of structure. In particular, the actual reptilian jaw-joint itself was relatively as powerful in the mammal, *Morganucodon*, as it was in the reptile *Cynognathus*. This was quite unexpected.[173]

Morganucodon and *Kuehneotherium* each possessed a full complement of the reptilian bones in its lower jaw.

Under the heading of "Mesozoic Mammals," Kemp states:

> *Kuehneotherium* held a pivotal role in the development of modern ideas about the evolution of mammals because of its interpretation as the most plesiomorphic holotherian mammal. However, there are some doubts about its phylogenetic position (Kielan-Jaworowska *et al.* 2004).[174]

Cynognathus ("dog jaws" of the lower Triassic. A wolf-sized therapsid, cynodont carnivore)

> Hunt's Claim in her Transitional Fossils FAQ: Cynodont theriodonts (e.g. *Cynognathus*)—very mammal-like reptiles. Or is that reptile-like mammals? Highly differentiated teeth (a classic mammalian feature), with accessory cusps on cheek teeth; strongly differentiated vertebral column

(with distinct types of vertebrae for the neck, chest, abdomen, pelvis, and tail—very mammalian), mammalian scapula, mammalian limbs, mammalian digits (e.g. reduction of number of bones in the first digit). But, still has unmistakably reptilian jaw joint.

1. What the fossil evidence actually shows: A spinal column with lumbar, dorsal, and cervical sections, and peg-like incisors in a rather large skull. The other teeth were well-designed (specialized). The elbow pointed backward and the knee pointed forward.
2. The evolutionary interpretation: Darwinists consider this extinct tetrapod as displaying characteristics intermediate between the mammals and reptiles.
3. The non-Darwinian interpretation: This is not a transition. Cynodonts "are believed to be the direct ancestors of mammals."[175] "Although there is still substantial disagreement regarding the phylogenetic relationships among the advanced cynodont clades, mammalian monophyly is well supported"[176]

There is no mention of *Cynognathus* in Allaby. Benton states that *Cynognathus* is a member of a clade of herbivorous forms, but "the cheek teeth show wear from processing meat."[177]

Pelycosaur synapsids ("mammal-like reptiles")

Hunt's Claim in her Transitional Fossils FAQ: Pelycosaur synapsids—classic reptilian skeleton, intermediate between the cotylosaurs (the earliest reptiles) and the therapsids.

1. What the fossil evidence actually shows: Synapsids have differentiated incisors, canines, and molars. Some synapsids had multiple jaw bones.
2. The evolutionary interpretation: Evolutionists suggest the pelycosaur synapsid skeleton is an intermediate between the therapsids and the cotylosaurs.
3. The non-Darwinian interpretation: This is not a transition. No one knows how or why synapsids supposedly evolved into mammals, but Darwinists assume various synapsid groups are somehow connected evolutionarily. Paleontology has shown there are significant differences between the early pelycosaurs and later therapsids. The required transitional forms have not been found.

Therapsids ("mammal-like reptiles" of the Permian and Triassic)

Hunt's Claim in her Transitional Fossils FAQ: Therapsids—the numerous therapsid fossils show gradual transitions from reptilian features to mammalian features. For example: the hard palate forms, the teeth differentiate, the occipital condyle on the base of the skull doubles, the ribs become restricted to the chest instead of extending down the whole body, the legs become "pulled in" instead of sprawled out, the ilium [*sic*] (major bone of the hip) expands forward.

Note: *Dimetrodon* was previously discussed in these pages as a pelycosaurian reptile, and the pelycosaur group is treated as synapsid mammal-like reptiles.

1. What the fossil evidence actually shows: Cynodonts were mammal-like reptiles that were dog-like in their overall structure and size.

   > The great majority of the characters found in therapsids are uninformative, being either plesiomorphic for Therapsida, or autapomorphic for the individual taxa. Although this may be the best cladogram, and therefore the basis of the best classification available, it does not inspire great confidence in its truth.[178]

2. The evolutionary interpretation: Therocephalians were formerly called "mammal-like reptiles," however, Benton is cautious, saying only that the palate of *Bauria* (Early Triassic) "is like the secondary palate of mammals" and "another *superficially* mammalian character in the loss of the bar of bone between the orbit and temporal fenestra."[179] In fact, some secular scientists maintain convergent evolution (which is not a scientific explanation, but an escape) must be called upon to "explain" other similarities therocephalians have with mammals.

3. The non-Darwinian interpretation: This is not a transition. Allaby calls the therapsids "an order of reptiles."[180] No fossils have been found that would give a credible line of descent from "early" pelycosaurids to "later" therapsids. "The transition between pelycosaurs and therapsids has not been documented."[181] Benton states "Both groups [pelycosaurs and Therapsida] together were formerly sometimes called 'mammal-like reptiles', although it should be noted that they neither are nor derive from reptiles."[182]

Tritylodont theriodonts (e.g., Tritylodon)

> Hunt's Claim in her Transitional Fossils FAQ: Tritilodont [*sic*] theriodonts (e.g. *Tritylodon, Bienotherium*)—skull even more mammalian (e.g. advanced zygomatic arches). Still has reptilian jaw joint.

1. What the fossil evidence actually shows: Decades ago, *Tritylodon* was discovered in South Africa (Upper Triassic beds). They were small animals with large zygomatic arches, unique tooth pattern (e.g., seven square-shaped cheek teeth on each side), and an elevated sagittal crest. More recently, a number of complete fossils of tritylodonts have been discovered.

2. The evolutionary interpretation: Evolutionists call these animals therapsids (reptiles) and believe they lived during the late Triassic. "Nevertheless, the old reptilian bones were still participants in the articulation and, therefore, the tritylodonts technically may be regarded as reptiles."[183]

3. The non-Darwinian interpretation: This was not a transition.

> Tritylodontids *are no more closely related to mammals than other derived cynodonts*, such as tritheledontids, brasilodontids, and dromatheriids.[184]
>
> Relationships among these three [Tritheledontidae, Tritylodontidae, Mammalia] mammaliamorph taxa are unresolved.[185]

Tritylodonts are not mentioned in Allaby.

Diarthrognathus (ictidosaur)

> Hunt's Claim in her Transitional Fossils FAQ: Ictidosaur theriodonts (e.g. *Diarthrognathus*)—has all the mammalian features of the tritilodonts, and has a double jaw joint; both the reptilian jaw joint and the mammalian jaw joint were present, side-by-side, in *Diarthrognathus*'s skull. A really stunning transitional fossil.

1. What the fossil evidence actually shows: Fossils of *Diarthrognathus* have been found in South Africa. Kemp shows a lower jaw (5 cm long) of *Diarthrognathus* in medial view,[186] as does Stahl.[187]

2. The evolutionary interpretation: Allaby states that *Diarthrognathus broomi* "is a species of mammal-like reptiles, recorded from the Triassic, that is one of the closest to mammalian ancestry. It is assigned to the Ictidosauria and is characterized by a number of advanced cranial features, including the co-occurrence of the older reptilian (quadrate-articular) and newer mammalian (squamosal-dentary) jaw joints."[188] Colbert says, "Here we see examples of the gradual transition from reptile to mammal. *Diarthrognathus*, from the Upper Triassic of South Africa, is on the reptilian side of the line because, although it had the double jaw joint, the quadrate-articular articulation was still dominant."[189]

3. The non-Darwinian interpretation: This is not a transition. If this was "a really stunning transitional fossil," then Anderson and Sues, 2007 and Dial et al, 2015 would certainly have listed and discussed *Diarthrognathus* in detail. There is no evidence of reptilian traits in *Diarthrognathus*. Benton states, "The tritheledonts are rather poorly known group of small animals that are mammal-like in many respects (Kemp, 1982; Martinelli and Rougier, 2007).... The skeletons of tritheledonts and tritylodonts show many mammal-like features."[190] It is significant Benton makes no mention of *Pachygenelus*, a tritheledontid cynodont. Stahl documents the animated discussions regarding the placement of *Diarthrognathus*.[191] (E.g., in the late 1960s, Jenkins, Hopson, and Crompton formed a new opinion of Mammalia origin that excluded *Diarthrognathus* based on its dentition.)

Monotreme, Marsupial, and Placental Mammals

There are three majors groups of milk-producing animals with hair (i.e., mammals): egg-laying monotremes (platypus and echidna), pouched marsupials (kangaroo, koala, etc.), and placentals (horses, dogs, cats, etc.), whose babies are nourished through the placenta in the womb. Marsupials and placental mammals apparently diverged from an unknown common ancestor in the Cretaceous. The common belief that these three are listed in order of evolutionary development is contradicted by stratigraphic position, morphology, and molecular data.

> While the origin of mammals has provided one set of puzzles, the diversification of mammals is also ripe with questions.[192]

In view of monotremes' reptilian affinities, they are thought by evolutionists to represent a separate and direct line of descent from the earliest Mesozoic animals, possibly the Docodonta, independent of the line leading to other mammals, but the dental features of *Steropodon* are considered by some to point to affinities with the Eupantotheria.

Under the title "Australosphenida and the mystery of the Monotremata," Kemp stated:

> The greatest mystery of all concerning mammalian evolution stretches back for 200 years: the question of what exactly the monotreme mammals are, and how they relate phylogenetically to therians (Musser and Archer 1998).[193]

Other evolutionists express the same confusion over the placement of monotremes:

> The monotremes *may have* had their origin in docodont ancestors, in turn derived from morganucodant-like progenitors.[194]
>
> Molecular phylogenetic data recently shook up the traditional understanding of the ordinal relationships among the birds, as well as those of the placental mammals. This type of data also questioned the previous hypothesis of the evolutionary relationships among the three major groups of mammals, the monotremes, marsupials, and placental mammals.[195]

Eupantotheres (e.g., Amphitherium)

> Hunt's Claim in her Transitional Fossils FAQ: Eupantotheres (e.g. *Amphitherium*)—these mammals begin to show the complex molar cusp patterns characteristic of modern marsupials and eutherians (placental mammals). Mammalian jaw joint.

1. What the fossil evidence actually shows: Some molars have been found and a mandible. Allaby states that the Eupanotheria are "known from the Jurassic of N. America and Europe and in its earliest, Middle Jurassic, form (*Amphitherium*) only from the lower jaw." The jaw identifies it as mammalian, and *possibly* ancestral to later pantotheres."[196]
2. The evolutionary interpretation: Eupanotheria are "Middle Jurassic" mammals.[197]
3. The non-Darwinian interpretation: This is not a transition.

> These orders of very ancient mammals, *some of which seem to be quite unrelated to the others*, are an indication that the transition from the reptilian to the mammalian stage of evolutionary development *most probably* took place along a broad front of adaptive radiation. The interrelationships of these orders of primitive mammals have been and still are *a subject of much dispute.*[198]

Benton and Stahl do not address Eupantotheria or the Amphitheriids. Kemp states:

> To this point in the story of the origin of tribosphenidans, the timing of the appearance in the fossil record of the different degrees of expression of the tribosphenic molar tooth has agreed with simple evolutionary expectations, from the Middle Jurassic *Amphitherium*, through the Late Jurassic *Peramus*, to the Early Cretaceous aegialodontids and the fully expressed version in the various younger Cretaceous lineages. *While not to be read literally as a sequence of ancestor-descendants*, these

forms do illustrate an approximate morphocline.[199]

In his mammalian diversification article in Anderson and Sues, Luo engages in speculation and inference regarding the meager fossil evidence of the Amphitheriids.[200]

Eutheria (placentals)

Allaby states the Eutheria is an "infraclass that includes all of the placental mammals and which *probably* arose during the Cretaceous."[201] Stahl describes the condition of the Late Cretaceous mammals:

> There is *a great gap* in the history of the mammals between the Forestburg years and Late Cretaceous time. *No fossil evidence at all* is available from the period during which, while the angiosperm flora supplanted the cycads and conifers, the older mammalian tribes failed and the newest therians proliferated. The late Cretaceous fauna, like the early one, is known from a relatively small number of localities. *Except for one tooth* from Europe, all the mammalian material that is definitely late Cretaceous in age comes from the upland deposits in Djadochta, Mongolia, and from North American sediments."[202]

Since Stahl's publication, Benton stated under "Eutheria: Jurassic and Cretaceous placentals," "The timing of eutherian origins was altered dramatically by the report of *Juramaia* from the Middle-Late Jurassic Tiaojishan Formation of China (Luo et al., 2011)....*Juramaia* shifts the origin of Eutheria, as well as all boreosphenidan nodes downwards by some 35 Myr."[203]

Benton goes on to discuss Placentalia:

> The origin of Placentalia, the crown group including modern placental mammals and their ancestors, is a much-discussed question, but there is little evidence for pre-Cenozoic placental mammal fossils (Wible et al., 2007, 2009).[204]

Luo in Anderson and Sues calls it a "basalmost eutherian."[205] But there are some problems. This would mean the first placental mammals must have existed even earlier, pushing one lineage back against the others. There also remains a "50 million year" gap in the Darwinian model that is essentially fossil-free.

Finally, Luo makes a statement that sounds very un-Darwinian: "Early mammalian evolution can be characterized as having a series of explosive events with modest diversity."[206] This sounds very much like the creation model—a sudden appearance followed by minor variation. Luo went on to say in the same paper:

> Notwithstanding these uncertainties, it is safe to conclude that all known Early Cretaceous eutherians (*Prokennalestes*, *Eomaia*, *Murtoilestes*, *Montanalestes*) and many Late Cretaceous eutherians (asioryctitherians and several other taxa) represent two successive episodes of diversification before the origin of crown Placentalia and cannot be directly related to any superorders of placentals. Early Cretaceous eutherians are evolutionary dead ends with respect to most of the Late Cretaceous eutherians, and many late Cretaceous eutherians (with the exception of zalambdalestids and zhelestids) are dead ends with respect to crown placentals.[207]

Proteutherians

> Hunt's Claim in her Transitional Fossils FAQ: Proteutherians (e.g. *Zalambdalestes*)—small, early insectivores with molars intermediate between eupantothere molars and modern eutherian molars.

1. What the fossil evidence actually shows: Colbert states that "[*Zalambdalestes*] was a small mammal, with a low skull less than 5 cm (2 in.) in length. There was a pair of enlarged, piercing incisors in the upper and lower jaws...the molars were tribosphenic, with the paracone and the metacone of the upper teeth widely separated on the outer border of each tooth."[208] Benton calls the Zalambdalestidae a non-placental eutherian family and gives a description of this "agile hedge-hog-sized animal with a long-snouted skull." He states, "There are typical numbers of teeth for a placental, four premolars and three molars. The molars are broad and they lack the

specializations of marsupial molars."[209]

2. The evolutionary interpretation: Allaby states, "There were many evolutionary side-branches which became extinct, but among successful proteutherians some *may be* ancestral to rodents and others *may be* ancestral to primates. Proteutherians are often excluded from the Insectivora or even divided into two or more separate orders."[210]

Stahl is again cautious: "A third line of placental mammals was represented by *Zalambdalestes*, an animal that *seems to have been* a rather specialized insectivore but which *might possibly have* been close to the ancestry of the rabbits."[211]

3. The non-Darwinian interpretation: This is not a transition. Many eutheria have difficult taxonomic histories that are not fully understood. Colbert admits that "the Cretaceous eutherians were rather similar to each other, as far as we can determine from their all too scanty remains."[212] In viewing the dentition of *Zalambdalestes*, Colbert says the "molars *appear to be* intermediate in structure between the eupantothere molar and the molar of the more advanced placentals."[213] Benton was refreshingly candid, stating:

> As noted, zhelestids had been classed as early ungulates, and the other Late Cretaceous eutherians have from time to time been assigned to modern placental clades. This is easy to do, with often incomplete fossils, and an optimistic desire to find the "oldest primate" or the "oldest rodent" for example.[214]

Bats

> Hunt's Claim: GAP: One of the least understood groups of modern mammals—there are no known bat fossils from the entire Paleocene. The first known fossil bat, *Icaronycteris*, is from the (later) Eocene, and it was already a fully flying animal very similar to modern bats. It did still have a few "primitive" features, though (unfused & unkeeled sternum, several teeth that modern bats have lost, etc.).

Chiroptera may be "one of the least understood groups of modern mammals," according to Ms. Hunt. The molecular phylogeny of these flying mammals continues to be controversial.

> Bats probably originated at a relatively early date, and they must have experienced an initial stage of very rapid evolution....there are no known intermediate stages between bats and insectivores.[215]

> Two particularly intensively studied early bats come from the early Eocene of Wyoming, *Icaronycteris* and *Onychonycteris*. These already show all the key microchiropteran features....Interestingly, in a species-level gene tree of bats, Agnarsson *et al* (2011) found mixed results, with most analyses supporting the Yin/Yang tree, but some retrieving the older Mega/Micro division of bats; Tsagkogeorga *et al* (2013) find the Yin/Yang split. This is an interesting case that shows the importance (and difficulty) of determining true phylogenies and their implications on evolutionary assumptions.[216]

> The first undoubted bats are preserved in middle Eocene deposits in both Europe and N. America.[217]

Any discussion of the massive order of bats (Chiroptera) is mysteriously missing in Anderson and Sues and Dial et al. (Indeed, if there was ever a major transition in vertebrate evolution—the bats would certainly qualify!)

The reason for the triumph of the artiodactyls over the other medium- and large-sized herbivores is not easy to explain.[218]

Next to the rodents, the artiodactyls are probably the most difficult mammals to classify.[219]

The dentition, and indeed the skull generally of *Diacodexis*, was little advanced from the "condylarth" grade and gives little away about the origin of the artiodactyls. An earlier proposal by Van Valen (1971) that they were related to the arctocyonid "condylarths" has received little support (Rose 1987; Prothero et al. 1988; Thewissen and Domning 1992).[220]

Diacodexis—the basal Eocene dichobunoid (artiodactyl)

Hunt's Claim in her Transitional Fossils FAQ: Transitional fossils from early hoofed animals to some of the artiodactyls (cloven-hoofed animals): Dichobunoids, e.g. *Diacodexis*, transitional between condylarths and all the artiodactyls (cloven-hoofed animals). Very condylarth-like but with a notably artiodactyl-like ankle.

1. What the fossil evidence actually shows: This North American creature was about the size of a rabbit with elongated limbs (longer than the condylarths) and lived during the early Eocene. The feet were elongated, and the third and fourth toes bore most of the weight.

   Diacodexis shows unique artiodactyl characters in the skull.[221]

2. The evolutionary interpretation: Evolutionists compare teeth of *Diacodexis* and *Chriacus* to make their case that this is a transitional form.

3. The non-Darwinian interpretation: This is not a transition. Comparing tooth similarities does not constitute a transitional form. *Chriacus* was an unspecialized condylarth. Creationists maintain there is no transitional evidence between *Chriacus* and *Diacodexis*.

Colbert states, "It is *very probable* that the early dichobunoids were ancestral to other nonruminant groups that arose during Eocene or early Oligocene times."[222] This is quite vague (i.e., "during Eocene or early Oligocene times" spans millions of years) and Colbert does not refer to these creatures as transitional. Allaby states these animals were small and primitive and showed "considerable variation" and have more recently "been regarded as a distinct artiodactyl group of subordinal status, the Palaeodonta. They were niether true swine nor true ruminants."[223]

Suiformes (hippos and pigs)

Hunt's Claim in her Transitional Fossils FAQ: Artiodactyls (cloven-hoofed animals) are represented among the transitional forms in the evolution of mammals (see above claim).

1. What the fossil evidence actually shows: These nonruminant artiodactyls first appeared during the Eocene.[224]

   The fossil remains of these animals [artiodactyls] are so numerous that no one has yet been able to review them all and draw a clear picture of the radiation of the modern lineages from their common stem.[225]

2. The evolutionary interpretation: The artiodactyls "are descended from the Condylarthra, and underwent a spectacular burst of adaptive radiation in Eocene and early Oligocene times."[226] "The first pigs and peccaries were of early Oligocene age."[227]

3. The non-Darwinian interpretation: This is not a transition.

   The entelodonts did have smaller relatives which spread as widely as they, and it is from these animals that paleontologists *believe* the modern pigs *may have come.*[228]

   Hippos date back to the early Miocene of Uganda and Kenya (Orliac *et al.*, 2010).[229]

Perhaps it is significant that there are no traces of hippopotamuses in the fossil record until late Miocene or early Pliocene times.[230]

The evolutionists' problem of what the fossils say (anatomy) and what the biological molecules say (molecular biology) is graphically seen here:

> The origin of late Neogene Hippopotamidae (Artiodactyla) involves one of the most serious conflicts between comparative anatomy and molecular biology: is Artiodactyla paraphyletic? Molecular comparisons indicate that Cetacea should be the modern sister group of hippos. This finding implies the existence of a fossil lineage linking cetaceans (first known in the early Eocene) to hippos (first known in the middle Miocene). The relationships of hippos within Artiodactyla are challenging, and the immediate affinities of Hippopotamidae have been studied by biologists for almost two centuries without resolution."[231]

Anthracotherium/anthracotherioids

> Hunt's Claim: *Anthracotherium* and later anthracotheriids (late Eocene)—A group of heavy artiodactyls that started out dog-size and increased to be hippo-size. Later species became amphibious with hippo-like teeth. Led to the modern hippos in the early Miocene, 18 Ma.

1. What the fossil evidence actually shows: Complete skeletons have been recovered of, for example, *Bothriodon*.
2. The evolutionary interpretation: The Anthracotheriidae were thought to be amphibious and therefore the ancestors of the hippopotami.
3. The non-Darwinian interpretation: This is not a transition. Evolutionists can only say some characters of the anthracotheres skeleton suggest they are related to hippos. However, anthracotherioids evidently were always anthracotherioids, changing only in size. Benton states, "The first anthracotheriids were small, but later ones became as large as pigmy hippos."[232] Stahl is diffident: "A branch of the family which migrated to North America in the Oligocene survived only briefly, but *it seems* that the anthracotheres *may have* left descendants in the Old World in the form of hippopotamuses."[233]

Propalaeochoerus

> Hunt's Claim: *Propalaeochoerus* or a similar cebochoerid/choeropotamid (late Eocene)—Primitive piglike artiodactyls derived from the helohyids.

1. What the fossil evidence actually shows: Complete fossils of this "fairly small artiodactyl"[234] were found in Europe. It had four-toed feet and a low, long skull. The teeth, including canines, were well-developed.
2. The evolutionary interpretation: *Propalaeochoerus* is seen by some evolutionists as being primitive and only pig-like.
3. The non-Darwinian interpretation: This is not a transition. Colbert makes no mention of *Propalaeochoerus* being a transition, saying it was "one of the first pigs."[235] On the following page, Colbert states, "The resemblances between the first known peccaries and the first known pigs are very striking."[236]

To assert that *Propalaeochoerus* is "derived from the helohyids" is not based on physical evidence but only on convictions that evolution occurred. *Propalaeochoerus* is not mentioned in Stahl, Allaby, Kemp, or Benton.

Paleochoerus (the first admitted 100 percent pig)

> Hunt's Claim: *Paleochoerus* (early Oligocene, 38 Ma)—First known true pig, apparently ancestral to all modern pigs. Pigs on the whole are still rather primitive artiodactyls; they lost the first toe on the forefoot and have long curving canines, but have very few other skeletal changes and still have low-cusped teeth. The main changes are a great lengthening of the skull & development of curving side tusks. These changes are seen *Hyotherium* (early Miocene), probably ancestral to the modern

pig *Sus* and other genera.

1. What the fossil evidence actually shows: Fossil evidence of *Paleochoerus* is well-documented.
2. The evolutionary interpretation: Some evolutionists feel this is an ancestor to modern swine (*Sus*).
3. The non-Darwinian interpretation: This is not a transition. Fossil evidence points to *Paleochoerus* as being 100 percent pig. Ms. Hunt can state only that it is "*apparently* ancestral to all modern pigs." *Paleochoerus* is not mentioned in Allaby, Colbert, or Benton. Stahl states, "Once the first of the modern pigs appeared in the Oligocene, all other pig-like animals except the amphibious anthracotheres vanished from the Old World."[237]

Perchoerus (an early peccary of the Oligocene age)

Hunt's Claim: *Perchoerus* (early Oligocene)—The first known peccary.

1. What the fossil evidence actually shows: Fossils show it had long canines.[238]
2. The evolutionary interpretation: Peccaries are of the family Tayassuidae, "omnivorous animals which are related to pigs but evolutionarily distinct from them for much of the Cenozoic. The oldest fossil peccary comes from the Oligocene of N. America." [239]
3. The non-Darwinian interpretation: This is not a transition. *Perchoerus* is a peccary variety. Colbert calls it "one of the early peccaries."[240] Benton calls *Perchoerus* an early peccary from the Oligocene of North America.[241] Stahl says, "The entelodonts did have smaller relatives which spread as widely as they, and it is from these animals that paleontologists *believe* the modern pigs *may have* come."[242] *Perchoerus* is not mentioned in Allaby.

Hyotherium, mentioned above by Ms. Hunt as "probably ancestral to the modern pig," is not mentioned in Allaby, Benton, Stahl, or Colbert.

Protylopus

Hunt's Claim in her Transitional Fossils FAQ: *Protylopus*, a small, short-necked, four-toed animal, transitional between dichobunoids and early camels. From here the camel lineage goes through *Protomeryx*, *Procamelus*, *Pleauchenia*, *Lama* (which are still alive; these are the llamas) and finally *Camelus*, the modern camels.

1. What the fossil evidence actually shows: Camel fossils appear in the late Eocene. Partial fossils of *Poebrodon* (partial mandible and some upper teeth) have been found in North America.
2. The evolutionary interpretation: Early camels and dichobunoids are bridged by *Protylopus*.
3. The non-Darwinian interpretation: This is not a transition. No mention is made of *Protylopus* or its supposed transitional status in Stahl, Allaby, Anderson and Sues, Colbert, or Benton. Allaby states, "Camelids appear first as fossils in N. American rocks of upper Eocene age."[243] No mention is made of any transitional forms. What Ms. Hunt discusses is not the macroevolution of camels and llamas, but the minor variation within the kinds.

Ruminants (the Main Selenodont Group: Antelope, Deer, Cattle, Mouse Deer, and Sheep)

It seems ruminants have always been ruminants, according to the fossil record. Stahl states:

> The best representatives of the common stock from which the most advanced ruminants arose are, strangely enough, not fossils but living forms. The chevrotain, *Tragulus*, that lives in Asia and its African cousin, *Hyemoschus*, have preserved almost without change the traits that paleontologists presume are primitive for the Pecora [cattle, deer, giraffes, and chevrotains].[244]

Ms. Hunt admits in regard to the ruminants, "It's been very difficult to untangle the phylogeny [evolution] of this fantastically huge, diverse, and successful group of herbivores."

Selenodontia (cattle, camels, and deer)

> Hunt's Claim: *Hypertragulus*, *Indomeryx* or a similar hypertragulid (late Eocene)—Primitive ruminants with a tendency toward crescent ridges on teeth, high-crowned teeth, and loss of one cusp on the upper molars. Long-legged runners and bounders, with many primitive features, but with telltale transitional signs: Still 5 toes on front and 4 behind, but the side toes are now smaller. Fibula still present (primitive), but now partially fused at the ends with the tibia. Upper incisors still present, but now smaller. Upper canine still pointed, but now the lower canine is like an incisor. Ulna and radius fused (new feature). Postorbital bar incomplete (primitive feature). Two ankle bones fused (new feature). Mastoid bone exposed on the surface of the skull (primitive feature).

1. What the fossil evidence actually shows: *Hypertragulus* fossils have been found in North America.
2. The evolutionary interpretation: *Hypertragulus* has been "dated" to the Oligocene.
3. The non-Darwinian interpretation: This is not a transition. *Hypertragulus* "is a small, rabbit-sized animal," according to Benton, and is not referred to as a transition.[245] Allaby mentions the Hypertragulidae as "a family of primitive ruminants," but no mention is made of any transitional forms.[246] Colbert cautiously states, "The ancestry of the ruminants may be approximated by various traguloids, notably...*Hypertragulus* from the Oligocene."[247] *Hypertragulus* is not mentioned in Stahl.

Deer

> Hunt's Claim in her Transitional Fossils FAQ: *Archeomeryx*, a rabbit-sized, four-toed animal, transitional between the dichobunoids and the early deer. From here the deer lineage goes through *Eumeryx*, *Paleomeryx* and *Blastomeryx*, *Dicrocerus* (with antlers) and then a shmoo [*sic*] of successful groups that survive today as modern deer—muntjacs, cervines, white-tail relatives, moose, reindeer, etc., etc.

1. What the fossil evidence actually shows: Fossils that are 100 percent deer have been found.
2. The evolutionary interpretation: Most zoology textbooks list members of the order Artiodactyla as even-toed mammals (hippopotamus, deer) that appeared in Eurasia in the Miocene and early Pliocene. By the end of the end of the Pliocene, they had radiated widely.
3. The non-Darwinian interpretation: This is not a transition. The *Cervidae* first appeared in the fossil record *as Cervidae*. The so-called "Irish elk" *Megaceros giganteus* (which is neither Irish, nor a true elk) was very much a cervid. It would seem these deer were already distinct and diverse in the earlier Oligocene. Stahl states, "The deer that appeared at the end of the Oligocene were already a diverse and widespread group."[248]

Hunt says *Archeomeryx* is "transitional," whereas Colbert uses the more cautious term "approximated" in regard to supposed ancestry.[249] Colbert states that the deer "seems" to have arisen from the Oligocene and that its ancestry can only be "approximated" via various traguloids (e.g., *Tragulus*—a living fossil chevrotain called the "mouse deer" in the Orient).

Rodents

> Hunt's Claim: Lagomorphs and rodents are two modern orders that look superficially similar but have long been thought to be unrelated. Until recently, the origins of both groups were a mystery. They popped into the late Paleocene fossil record fully formed—in North America & Europe, that is. New discoveries of earlier fossils from previously unstudied deposits in *Asia* have finally revealed the probable ancestors of both rodents and lagomorphs—surprise, they're related after all. (See Chuankuei-Li et al., 1987.)[250]

It is significant that Ms. Hunt admitted that rodents and lagomorphs "popped into the late Paleocene fossil record fully formed—in North America & Europe." Non-Darwinists certainly agree to this sudden and un-evolutionary appearance. Under the Lagomorph heading below, we will see if professionals agree with Ms. Hunt's claim for a lagomorph-rodent link. The work of Chuankuei-Li et al, cited by Hunt as supposedly revealing "the probable ancestors of both [Asian] rodents and lagomorphs," is not mentioned in Colbert or Benton.

There are about 1,700 species of rodents. They suddenly appear in the fossil record—*as rodents*. There is no trace of any ancestry linking them to non-rodent ancestors. The fossils are reasonably common, but they show *no* evolutionary progression. M. J. Mason states, "Homoplasy is rife among the various rodent clades."[251] Hautier and Cox candidly state:

> Indeed from a quick look at the fossil record, it is easy to get the impression that rodents have always been rodents.[252]

The latest evolutionary candidates for rodent ancestors are the Eurymylidae, from the early Tertiary of Asia. Recently, evolutionists suggested that a creature called *Heomys* may be the possible ancestor of rodents, although it is too advanced and its appearance is too late to be an ancestor. Other eurymylids such as *Matutinia*, *Rhombomylus,* and *Eurymylus* are a side branch and not directly ancestral to rodents.[253]

Squirrels (sciuromorphs)

Squirrels appear in the fossil record (middle Tertiary) as 100 percent squirrels. The oldest squirrel fossil, *Douglassciurus,* is from the Late Eocene but is still a nearly-modern tree squirrel. As Kathleen Hunt has said, the modern squirrel genus "arose in the Miocene and has not changed since then."

Squirrels (i.e., *Sciurus*) have always been squirrels with no sign of a macroevolutionary past. However, evolutionists maintain that *Paramys* (Eocene ischyromid) is the evolutionary ancestor of squirrels (as well as beaver, jumping mice, Old and New World porcupines, and other mammals!). Benton says the oldest squirrels are from the early Eocene age—evidently as 100 percent squirrels.[254] Roth and Mercer state, "Other themes are conservativism, correlated, and convergent evolution that have made it difficult to reconstruct phylogeny within the Sciuridae using morphological characters."[255]

Palaeocastor (an Oligocene/Miocene beaver-like rodent)

> Hunt's Claim: *Paleocastor* (Oligocene)—Early beaver. A burrower, not yet aquatic. From here the beaver lineage became increasingly aquatic. Modern beavers appear in the Pleistocene.

1. What the fossil evidence actually shows: Skeletons of *Palaeocastor* "have been found associated with the natural casts of their burrows."[256] (In 2006, an amazing animal called the "Jurassic Beaver" was discovered.[257] It was dated at "164 million years" with modern skin and fur structure.)
2. The evolutionary interpretation: *Paleocastor* lived "20 million years ago" as beavers.
3. The non-Darwinian interpretation: This is not a transition. No solid fossil evidence exists (i.e., a series of intermediate forms). *Palaeocastor* is not mentioned in Allaby, Kemp, Cox and Hautier, or Stahl, and Colbert does not call *Palaeocastor* a transition. From what can be determined, this was a terrestrial beaver—but 100 percent beaver nonetheless.

Rats

There are Old World (Muridae) and New World (Cricetidae) rats. The Muridae suddenly appear in the Oligocene and the Cricetids during the Miocene.

Lagomorphs (Pikas, Hares, and Rabbits)

Fossils of 100 percent lagomorphs have been found in sedimentary rock. In addition, Allaby states in regard to Glires:

> The cohort includes the orders Rodentia and Lagomorpha presumed to have diverged from the insectivorous eutherian stock during the Cretaceous and subsequently differentiated into rodent and lagomorph forms; but similarities between the two orders may be superficial, implying no close relationship, so their inclusion in a single cohort is somewhat arbitrary.[258]

Under "The Cretaceous: origin and radiation [of placentals]," Kemp states:

> This marked and consistent lack of agreement between the pictures based respectively on the fossils and the molecular evidence has led to considerable controversy over attempts at a resolution (Foote et al. 1999; Archibald and Deutschman 2001).[259]

A diagram in Colbert clearly shows rodents as rodents and lagomorphs as lagomorphs[260]—just as the creation model states. Indeed, evolutionist M. L. Weston stated in the *Canadian Encyclopedia*, "The origin of the lagomorphs is uncertain."[261] Allaby writes, "The lagomorphs are *believed* to have diverged from a primitive eutherian stock at the same time as, or soon after, the rodents."[262]

We find that although evolution theory states most placental orders (including rodents) are related to the earliest insectivores, there's no trace of any ancestry linking rodents to non-rodent predecessors (e.g., early insectivores). Benton says, "The rabbit fossil record extends back to the early Eocene."[263]

The first rodent-like mammals (*Paramys*) that appear in the supposed Paleocene are already fully developed with incisors unique to rodents—as the creation model predicts. Rodent molars are designed to grind plant material—an ability that is unique from the early insectivores from whence they supposedly evolved. Benton writes, "Rodent phylogeny has been much discussed and there has been a broad range of viewpoints."[264] Hautier and Cox stated, "Although intensively studied, the phylogenetic relationships between the different groups of rodents have been a matter of debate for over 150 years."[265]

In March 2008, paleontologists announced a fossil find of a "53 million-year-old" rabbit foot bone that, according to evolutionary dating, indicated that true rabbits existed "20 million years" before evolutionists had previously thought.[266] Rabbits have always been rabbits, just as the creation model states.

Barunlestes (a Late Cretaceous placental mammal)

> Hunt's Claim: *Barunlestes*...the possible Asian rodent/lagomorph ancestor.

1. What the fossil evidence actually shows: This creature has the physical evidence of being a mouse.
2. The evolutionary interpretation: Some evolutionists speculate that *Barunlestes* was a possible Asian lagomorph/rodent ancestor.
3. The non-Darwinian interpretation: This is not a transition. *Barunlestes* is not mentioned in Allaby, Benton, Stahl, or Anderson and Sues. Ms. Hunt uses the word "possible." Very well, it is possible that *Barunlestes* is simply a contemporary of *Anagale* and *Heomys*, rather than an ancestral form. (See Benton, 2015, 361-362.)

Mimotoma

> Hunt's Claim: *Mimotoma* (Paleocene)—A rabbit-like animal, similar to *Barunlestes*, but with a *rabbit* dental formula, changes in the facial bones, and only one layer of enamel on the incisors (unlike the rodents). Like rabbits, it had two upper incisors, but the second incisor is still large and functional, while in modern rabbits it is tiny. Chuankuei-Li et al...think this is the actual ancestor of *Mimolagus*.[267]

1. What the fossil evidence actually shows: Some facial bones and teeth of *Mimotoma* have been found.

2. The evolutionary interpretation: Evolutionists Chuankuei-Li et al, cited by Ms. Hunt, suspect *Mimotoma* is ancestral to *Mimolagus*, although there is a serious dating discrepancy (i.e., an approximate "10 million year" gap).
3. The non-Darwinian interpretation: This is not a transition. *Mimotoma* is not even mentioned in Benton, Anderson and Sues, Allaby, or Stahl. Colbert describes the Mimotonida as "another poorly known extinct group related to the rabbits."[268] An argument could be made that *Mimotoma* is older than *Barunlestes*, or possibly they were contemporaries.

Mimolagus

> Hunt's Claim: *Mimolagus* (late Eocene)—Possesses several more lagomorph-like characters, such as a special enamel layer, possible double upper incisors, and large premolars.

1. What the fossil evidence actually shows: This is known only by a few teeth.
2. The evolutionary interpretation: According to Kathleen Hunt, it "possesses several more lagomorph-like characters, such as a special enamel layer, possible double upper incisors, and large premolars."
3. The non-Darwinian interpretation: This is not a transition. It is not mentioned in Colbert, Stahl, Allaby, Anderson and Sues, or Benton. There seems to be no evidential connection between *Mimolagus* and its supposed ancestor *Mimotoma*.

Lushilagus (a lagomorph)

> Hunt's Claim: *Lushilagus* (mid-late Eocene)—First true lagomorph. Teeth very similar to Mimotoma, and modern rabbit & hare teeth could easily have been derived from these teeth. After this, the first modern rabbits appeared in the Oligocene.

1. What the fossil evidence actually shows: Evidently, only teeth have been found.
2. The evolutionary interpretation: According to Ms. Hunt, this is the "first true lagomorph. Teeth very similar to *Mimotoma*, and modern rabbit and hare teeth could easily have been derived from these teeth."
3. The non-Darwinian interpretation: This is not a transition. The phrase "could easily have been derived" is not scientific—it is wishful evolutionary speculation. *Lushilagus* could not have evolved from *Mimolagus* due to the evolutionary dating discrepancies. *Lushilagus* is not mentioned in Colbert, Stahl, Allaby, Anderson and Sues, or Benton.

Prolagus (a lagomorph)

> Hunt's Claim: The mid-Tertiary lagomorph *Prolagus* shows a very nice "chronocline" (gradual change over time), grading from one species to the next. Gingerich (1977) says: "In *Prolagus* a very complete fossil record shows a remarkable but continuous and gradual reorganization of the premolar crown morphology in a single lineage."[269]

1. What the fossil evidence actually shows: The family Prolagidae is an extinct family of rabbits within the order of lagomorphs.
2. The evolutionary interpretation: In 1977, Gingerich cited evidence for *Prolagus* evolution regarding a portion of some of the teeth of "a single lineage."
3. The non-Darwinian interpretation: This is not a transition. *Prolagus* is not mentioned in Allaby, Stahl, Colbert, Benton, or Anderson and Sues, and therefore the supposed "very nice chronocline" is highly suspect—or the professionals simply don't think such a minor change ending in extinction merits attention.

Choosing to "major on the minors," Ms. Hunt makes another claim about rabbit variation:

> Lundelius et al. (1987) mention transitions in Pleistocene rabbits, particularly from *Nekrolagus*

> to *Sylvilagus*, and from *Pratilepus* to *Aluralagus*. Note that both these transitions cross genus lines. Also see the lagomorph paper in Chaline (1983). Some of these transitions were considered to be "sudden appearances" until the intervening fossils were studied, revealing numerous transitional individuals.[270]

The "transitions" mentioned in the above paragraph are much like the variation within the horse kinds. Benton makes no mention of these Pleistocene rabbits. Colbert mentions *Sylvilagus* only once—but not as a transition.[271]

Whale Evolution

> Hunt's Claim: The Order Perissodactyla (horses, etc.) and the Order Cetacea (whales) can both be traced back to early Eocene animals that looked only marginally different from each other, and didn't look *at all* like horses or whales.

There is an ongoing debate within the Darwinian camp regarding the supposed evolution of whales from non-whale ancestors. A team from Northeastern Ohio University headed by Hans Thewissen has one idea, but Canadian Jessica Theodor and Jonathan Geisler of Georgia Southern University have another.

Thewissen maintains that whales are more related to an extinct pig-like animal, known as *Indohyus* of India, and that the hippopotamus is more related to pigs—not whales. Theodor and Geisler state DNA phylogeny is important and such an evaluation places the hippo in the whale family tree. Thewissen and his co-workers do agree with sections of their opponents' Darwinian tree, but still maintain hippos are closer to true pigs. A University of Calgary news release reported:

> Geisler and Theodor argue that leaving out the DNA data not only ignores important information, it implies that the evolution of swimming evolved independently in hippos and whales, when it may have evolved only once in a common ancestor.[272]

In 2005, an article in the *Proceedings of the National Academy of Sciences* stated:

> The origin of late Neogene Hippopotamidae (Artiodactyla) involves one of the most serious conflicts between comparative anatomy and molecular biology: is Artiodactyla paraphyletic? Molecular comparisons indicate that Cetacea should be the modern sister group of hippos. This finding implies the existence of a fossil lineage linking cetaceans (first known in the early Eocene) to hippos (first known in the middle Miocene). The relationships of hippos within Artiodactyla are challenging, and the immediate affinities of Hippopotamidae have been studied by biologists for almost two centuries without resolution.[273]

Secular science has shown hippos couldn't possibly be the ancestors of whales because hippos don't appear in the fossil record until 35 million years (according to the evolutionary dates) after whales diverged from their supposed land-dwelling ancestors.

The "whale wars" continue while creationists maintain that whales have always been whales, as the fossil record shows. Evolutionary writings reflect the lack of a satisfactory evolutionary explanation:

> Looking at the great blue whale, 30 m long, or a fast-swimming dolphin, *it is hard to imagine* how they evolved from terrestrial mammal ancestors, and yet that is what happened. [274]
>
> Like the bats, the whales (using this term in a general and inclusive sense) appear suddenly in early Tertiary times, fully adapted by profound modifications of the basic mammalian structure for a highly specialized mode of life. Indeed the whales are even more isolated with relation to other mammals than the bats; they stand quite alone.[275]
>
> Regardless of the identity of the sister-group of cetaceans, the question of what makes a whale a whale is still unanswered.[276]

Some evolutionists acknowledge that selection doesn't seem up to the herculean task of evolving a whale in a short time period.

> But was the jump between *Ambulocetus* and *Protocetus* and that between *Protocetus* and *Basilosaurus* the result of gradual cumulative selection? Some argue that these great morphological changes appeared so rapidly in the fossil record (within less than 10 million years) that they could not possibly have arisen and been fixed by Darwinian selection.[277]

Explanations of how tetrapods supposedly evolved into cetaceans are reduced to the following fanciful story:

> The sea cows evolved forelimbs which could be used as paddles and they took to the sea.[278]

"The sea cows evolved" is not a scientific explanation.

Kentriodon

> Hunt's Claim in her Transitional Fossils FAQ: *Kentriodon*, an early toothed whale with whale-like teeth.

1. What the fossil evidence actually shows: Fossils of *Kentriodon* have been found. Benton shows the "telescoping of the skull elements in the dorsal view of the skull of *Kentriodon*."[279]
2. The evolutionary interpretation: Evolutionists consider *Kentriodon* an early toothed whale of the Early Miocene.
3. The non-Darwinian interpretation: This is not a transition. Benton states that *Kentriodon* is a Miocene *dolphin*,[280] not "an early toothed whale," as Kathleen Hunt says. Colbert calls *Kentriodon* "a common odontocete" but doesn't call it a transition.[281] *Kentriodon* is not mentioned in Allaby or Stahl.

Mesocetus

> Hunt's Claim in her Transitional Fossils FAQ: *Mesocetus*, an early whalebone whale. (Note: very rarely a modern whale is found with tiny hind legs, showing that some whales still retain the genes for making hind legs.)

1. What the fossil evidence actually shows: Fossils show this was a baleen whale variety (i.e., toothless filter feeder).
2. The evolutionary interpretation: Colbert states that *Mesocetus* was "typical" of the group of primitive mysticetes (baleen whales).[282]
3. The non-Darwinian interpretation: This is not a transition. *Mesocetus* is not mentioned in Benton, Allaby, or Stahl.

Evolutionists are unable to present a single ancestral species of mesonychids that would result in whales today. Allaby states only that it is "believed" that the Mesonychidae "show that modern ungulates and carnivores are descended from a common stock."[283] Evolutionist Robert Carroll said, "It is not possible to identify a sequence of mesonychids leading directly to whales."[284] The community of evolutionary paleontologists is divided on this issue, with a majority stating that whales did not come from the mesonychids and a vocal minority maintaining that mesonychids are the ancestors of whales.

Indohyus (known as "India's pig," a "48 million-year-old" artiodactyl)

> Claim: Although not listed in Kathleen Hunt's FAQ, *Indohyus* has been cited as an "intermediate step" in the evolution of whales.

1. What the fossil evidence actually shows: Hundreds of these racoon-sized fossils have been discovered in mudstone riverbeds in Kashmir, India. The creature may have looked like a small deer and was an artiodactyl (which includes giraffes, pigs, sheep, and hippos).
2. The evolutionary interpretation: Dated at 48 million years old, it is claimed to be a missing link in whale evolution and is considered a likely ancestor of cetaceans. Supposedly the early cetaceans and *Indohyus* (possibly belonging to the family of raoellids) descended from an unknown common ancestor.
3. The non-Darwinian interpretation: This is not a transition. This discovery challenges macroevolutionary ideas regarding whale evolution—i.e., the idea that cetaceans split from forebears that lived on land and returned to the seas in search of prey. Clearly, not all paleontologists are convinced that *Indohyus* is a transition.[285]

There is no compelling scientific reason to claim *Indohyus* was on its way to becoming a 100-foot blue whale. It should be noted that this "48 million-year-old" deer-like creature supposedly ate only plants, but it evolved into an aquatic carnivore.

Pakicetus ("earliest known whale")

> Hunt's Claim in her Transitional Fossils FAQ: *Pakicetus*—the oldest fossil whale known. Only the skull was found. It is a distinct whale skull, but with nostrils in the position of a land animal (tip of snout). The ears were partially modified for hearing under water. This fossil was found in association with fossils of land mammals, suggesting this early whale maybe could walk on land.

1. What the fossil evidence actually shows: In 1981 a very incomplete skull (13 in. long) of an otter-sized creature was found in Pakistan. The body skeleton was then unknown. It was called *Pakicetus inachus*. Fragmented portions of the skeleton were later found.
2. The evolutionary interpretation: Evolutionists title this the earliest known ("distinctly cetacean") whale, stating the few skull fragments were found in Eocene rock. The body length was estimated at over six feet. In 2001 more bones that supposedly belong to *Pakicetus* were discovered and discussed by whale expert Thewissen and colleagues in *Nature*.[286]
3. The non-Darwinian interpretation: This is not a transition. An artist's rendition of a complete *Pakicetus* swimming after fish was based only on evolutionary artistic license (imagination), since skull pieces were all that was found. Prestigious journals such as the *Journal of Geological Education*[287] and *Science*[288] enthusiastically embraced *Pakicetus*' transitional status. Professionals were much more cautious. Benton states, "*Pakicetus* from the early Eocene, *reconstructed* skull in lateral view and *tentative* life restoration."[289] He also states, "The skeleton of *Pakicetus* is incompletely known."[290]

Colbert shows a "restoration" of *Pakicetus* and its "possible appearance."[291] He says this creature "was fully aquatic,"[292] but this can only be assumed. Benton stated:

> The skeleton of *Pakicetus* is *incompletely known,* and an early tentative reconstruction showed a semi-aquatic coast-dwelling carnivore. This was debated, and some more terrestrial reconstructions were presented, on the basis of the extensive retention of adaptations such as the typical artiodactyl double-pulley astragalus. In reviewing the evidence Madar (2007) concluded that pakicetids were capable of walking on land, but not sustained running, and that they spent more time in the water, moving around by bottom walking, paddling, and undulatory swimming."[293]

An earlier commentary on the subject in *Nature* stated, "All the postcranial bones indicate that pakicetids were land mammals, and...indicate that the animals were runners, with only their feet touching the ground.[294] *Pakicetus* seems to be of the same variety as the less famous *Hapalodectes* (otter-like mesonychids).

To review, *Pakicetus* was, according to:

> Colbert—fully aquatic, swimming
> Benton—semi-aquatic, walking
> *Nature*—land mammals, running

Strangely, *Pakicetus* is not mentioned in Hickman et al or Allaby.

Basilosaurus

> Hunt's Claim in her Transitional Fossil FAQ: *Basilosaurus isis*—a recently discovered "legged" whale from the Eocene (after *Pakicetus*). Had hind feet with 3 toes and a tiny remnant of the 2nd toe (the big toe is totally missing). The legs were small and must have been useless for locomotion, but were specialized for swinging forward into a locked straddle position—probably an aid to copulation for this long-bodied, serpentine whale.

1. What the fossil evidence actually shows: A 60-foot (nearly 20 m)-long creature was recently discovered with a complete, well-formed hind limb, including a foot with three toes. It had a small head with teeth having a comb-like pattern, and a long, thin body with a flexible backbone. The nostrils were possibly halfway between the top of the head and the front of the face.
2. The evolutionary interpretation: Colbert classifies *Basilosaurus* as an early whale.[295] Benton does not call this a "legged" whale but "the first giant whale."[296] In regard to the recently discovered hind limbs, three evolutionists state, "Thus hind limbs of *Basilosaurus* are most plausibly interpreted as accessories facilitating reproduction."[297]
3. The non-Darwinian interpretation: This is not a transition. The Basilosaurinae was an isolated subfamily. Evolutionists supposed the hind limbs of this animal were suppressed due to their belief in evolution theory. These limbs very probably *de*-volved. Furthermore, to say that *Basilosaurus* was a "legged" whale is to assume the evolution of whales and impress this preconceived viewpoint on fossils that have been pronounced to be a missing link. It misses the point that these structures are designed for mating in water—not walking on land. This creature, however, "did not give rise to modern taxa," according to J. G. M. Thewissen, a leading promoter of whale evolution.[298]

Basilosaurus had nothing to do with modern whale origins. Its Eocene location and aquatic designs make it too early stratigraphically and too advanced morphologically to qualify as both an insectivore descendant and cetacean ancestor—unless miracles are not excluded.

Squalodonts (e.g., Prosqualodon)

> Hunt's Claim in her Transitional Fossils FAQ: Squalodonts (e.g., *Prosqualodon*)—whale-like skull with dorsal nostrils (blowhole), still with un-whale-like teeth.

1. What the fossil evidence actually shows: The teeth of this group resembled those of sharks. They had dorsal nostrils.
2. The evolutionary interpretation: *Prosqualodon* lived about "16 to 30 million years" ago in the Miocene and Oligocene. Evolutionists call this group primitive odontocetes (toothed whales). Colbert says these were "important whales during the Miocene epoch, but they did not survive long."[299] Evolution theory states *Prosqualodon* and *Agorophius* are evolutionarily related.
3. The non-Darwinian interpretation: This is not a transition. Squalodonts were an extinct supergroup of toothed whales and not transitional. *Prosqualodon* is not mentioned in Stahl, Allaby, or Benton. Both *Agorophius* and *Prosqualodon* are assigned to the late Oligocene, so they cannot be considered ancestral to each other.

Perissodactyls (Horses, Tapirs, and Rhinos)

Order Condylarthra is an extinct order of "primitive" ungulates (grazing, hoofed mammals). Barbra Stahl states rather cryptically:

> *Evidently* descended from *some* Paleocene condylarth stock, the two kinds of ungulates developed distinguishing characteristics but paralleled each other in many adaptive traits. They are segregated on the basis of the structure of their feet into two orders, Perissodactyla and Artiodactyla.[300]

A bizarre group of perissodactyls are the chalicotheres. Kemp states they "survived into the Pleistocene of Asia and Africa and indeed there have been occasional, unsubstantiated reports from Kenya of sightings of a living chalicotheres (Savage and Long, 1986)."[301] Allaby states the "condylarths appear to be transitional between insectivores and true ungulates."[302]

Horses

The "Timescale and Horse Family Tree" Ms. Hunt presents on her horse evolution page in the TalkOrigins Archive[303] is really an artifact of selection and is better interpreted as a creationist "thick bush" rather than an evolutionary "tree." In fact, evolutionist Benton states:

> Note also that the evolution in the horses is "bushy," especially in the Miocene, not "ladder"-like.[304]

Evolutionary stories regarding horse evolution are constantly changing. Colbert states, "When all fossils are taken into account, the history of horses in North America is seen to be anything but a simple progression along a single line of development."[305]

Research done in 2009 confirms Colbert's statement. Something close to a complete overhaul of "horse evolution" is required following the detailed research by a team of 22 international researchers led by Ludovic Orlando of the University of Lyon in France. They subjected broad comparisons of ancient DNA (aDNA) from fossil equids (including zebras, horses, and donkeys). The results were unbelievable. They found a very un-Darwinian burst of diversification within the horse group—much like the perplexing Cambrian explosion of life forms at the base of the "geologic column." The international researchers also determined that many specimens that were supposedly separate species are actually variations of the same species.

Their article in the *Proceedings of the National Academy of Sciences* begins:

> The rich fossil record of the family Equidae (Mammalia: Perissodactyla) over the past 55 MY [million years] has made it an icon for the patterns and processes of macroevolution. Despite this, many aspects of equid phylogenetic relationships and taxonomy remain unresolved. Recent genetic analyses of extinct equids have revealed unexpected evolutionary patterns and a need for major revisions at the generic, subgeneric, and species levels.[306]

A perissodactyl is an order of mammals that contains odd-toed ungulates, such as the one-toed horse. Clearly, for example, that is what the lightly built *Hyracotherium* (which existed "55 million years ago") was. The creature was designed to run on its thin legs. Orlando et al stated, "The original linear model of gradual modification of fox-sized animals (Hyracothere horses) to the modern forms has been replaced by a more complex tree, showing periods of explosive diversification and branch extinctions over 55 MY."[307]

Allometry (relative relationship between shape and size during a creature's growth) is seen, for example, in quantitative skull growth within the family Equidae. The changes within the various horse types could be accomplished by gene segregation. If this seems surprising, one should consider the various types of dogs that all belong to the genus *Canis*. Horses, with their "tremendous body-size plasticity," experience "explosive diversification" in the fossil record:

> The end of the Early Miocene (15–20 MYA) marks a particularly important transition, separating an initial phase of small leafy browsers from a second phase of more diverse animals, exhibiting tremendous body-size plasticity and modifications in tooth morphology. This explosive diversifica-

tion has been accompanied by several stages of geographic extension from North America to the rest of the New and Old Worlds.[308]

It is not surprising to find buried together fossils that were once thought to be different stages of "horse evolution." Horse fossils are in a reversed sequence in South America, with small forest browsers with multiple hooves on the top and large grassland grazers with one hoof on the bottom of sedimentary rock units. Creationists suggest the supposed "horse evolution series" can be viewed as an ecological representation (creation), and clearly is not descent-with-modification (vertical evolution).

Indeed, Colbert states, "There were several episodes of adaptive radiation" in North American horses.[309] But adaptive radiation is not macroevolution. A common example given by evolutionists of adaptive radiation is the formation of 13 "new species" of Darwin's finches on the Galapagos Islands. However, as admitted by evolutionist Jonathan Weiner, these new species of finch are able to interbreed.[310] Appealing to adaptive radiation in the horse series is not documentation of macroevolution, but variation of the created kind.

There are serious problems between the evolutionary interpretations of those who study fossils (paleontologists) and molecular biologists who analyze DNA data from an evolutionary perspective. The genetic data produced by Orlando et al have shown that equids that appear morphologically different are really just variations of the same kinds. The final paragraph of this revealing paper shows serious problems for the evolutionist in other areas:

> This pattern of taxonomic oversplitting does not appear to be restricted to equids but is widespread amongst other Quaternary megafauna [e.g., Late Pleistocene bison (49); Holarctic cave lions (50); New World brown bears (51), and ratite moas (52, 53)]. Together, these findings suggest that the morphological plasticity of large terrestrial vertebrates across space and time has generally been underestimated, opening the way to detailed studies of the environmental, ecological, and epigenetic factors involved. Interestingly, in this regard the human lineage shows a rich fossil record over the last 6 MY, spreading over seven possible genera and 22 species (54). The exact number of taxonomic groups that should be recognized is still debated, even within our own genus (55), and in this context it is pertinent to consider the degree of taxonomic oversplitting, from species to generic levels, that aDNA has revealed amongst Late Pleistocene equids and other megafauna.[311]

Tapirs (odd-toed ungulates)

Hunt's Claims in her Transitional Fossils FAQ:

Hyrachyids—transitional from perissodactyl-like condylarths to tapirs.

Heptodonts, e.g. *Lophiodont*—a small horse-like tapir, transitional to modern tapirs.

Protapirus—a probable descendent of *Lophiodont*, much like modern tapirs but without the flexible snout.

Miotapirus—an almost-modern tapir with a flexible snout, transitional between.

Protapirus and the modern *Tapirus*.

1. What the fossil evidence actually shows: Tapirs have a poor fossil record. The fossil evidence that has been found (see point 3 below) shows tapirs to be tapirs. The above claims by Ms. Hunt suggest relatively minor variation.
2. The evolutionary interpretation: Colbert states, "In Oligocene times *Protapirus* appeared as a probable descendant of *Heptodon,*" and after briefly discussing the nasal area, he goes on to say, "From here on, the evolution of the tapirs was comparatively simple, and involved mainly a certain degree of size increase."[312] Allaby states the Lophiodontidae (Ceratomorpha) "are known only from Eocene sediments and were closely related to the tapirs."[313]
3. The non-Darwinian interpretation: These are not transitions. The supposed evolution of tapirs is very confusing, with whole species being described from a fossilized tooth. Ms. Hunt lists tapir varieties and then states

NOTES

Foreword

1. Glenister, B. F. and B. G. Witzke, 1983, *Did the Devil Make Darwin Do It?,* D. B. Wilson, ed., Ames, IA: State University Press, 58.
2. Futuyma, D. J., 1983, *Science on Trial*, New York: Pantheon Books, 197.
3. Lewontin, R., 1983, In the Introduction to *Scientists Confront Creationism,* L. R. Godfrey, ed., New York: W.W. Norton and Co., xxvi.
4. Strahler, A. N., 1987, *Science and Earth History – The Evolution/ Creation Controversy*, Buffalo: Prometheus Books, 408.

Introduction

1. E. O. Wilson, 1982, Toward a Humanistic Biology, *The Humanist*, 42: 40.
2. R. L. Numbers, 1992, *The Creationists*, New York: Alfred A. Knopf, xvi.
3. M. J. Arct, 1991, Dendroecology in the Fossil Forests of the Specimen Creek Area, Yellowstone National Park, Ph.D. dissertation, Loma Linda University, Loma Linda, CA. See also his M.S. thesis, Dendoecology in Yellowstone Fossil Forests (1979).

Chapter 1

1. M. Ruse, Saving Darwin from the Darwinians, *National Post*, May 13, 2000, B-3.
2. R. Lewontin, 1997, Billions and Billions of Demons: Review of *The Demon-Haunted World* by Carl Sagan, *The New York Review of Books*, 44 (1): 31.

Chapter 2

1. S. J. Gould, 1977, Evolution's Erratic Pace, *Natural History,* 86 (6).
2. Gould, 1977, 22-30.
3. S. Stanley, 1977, *Macroevolution*, San Francisco, CA: Freeman, 39.
4. H. Powell, Forest songsters evolved in an early burst of innovation, Cornell University news release, July 9, 2008, reporting on research published in D. L. Rabosky and I. J. Lovette, 2008, Density-dependent diversification in North American wood warblers, *Proceedings of the Royal Society B*, 275 (1649): 2363-2371.
5. M. Ridley, 1981, Who doubts evolution? *New Scientist,* 90: 832.
6. E. V. Koonin, 2007, The Biological Big Bang model for the major transitions in evolution, *Biology Direct*, 2: 21.
7. P. C. J. Donoghue, 2007, Palaeontology: Embryonic identity crisis, *Nature*, 445 (7124): 155.
8. S. Mazur, Altenberg! The Woodstock of Evolution?, posted on suzanmazur.com March 4, 2008.
9. S. J. Gould, 1991, Not Necessarily a Wing, *Bully for Brontosaurus*, New York: W. W. Norton & Co., 139-151 (originally published October 1985 as Not Necessarily a Wing, *Natural History*, 94: 12-25).

Chapter 3

1. D. Futuyma, 1983, *Science on Trial*, New York: Pantheon Books, 197.
2. E. R. Noble, G. A Noble, G. A. Schad, and A. J. MacInnes, 1989, *Parasitology: The Biology of Animal Parasites*, 6th ed., Philadelphia, PA: Lea & Febiger, 516.

Chapter 4

1. See J. D. Morris, 2007, *The Young Earth*, Green Forest, AR: Master Books; and L. Vardiman, A. A. Snelling, and E. F. Chaffin, eds., 2000, *Radioisotopes and the Age of the Earth*, vol. 1, El Cajon, CA: Institute for Creation Research, and St. Joseph, MO: Creation Research Society.

Chapter 5

1. G. Englmann and G. Callison, 1998, Mammalian faunas of the Morrison Formation, *Modern Geology*, 23: 343-380.
2. C. Turner and F. Peterson, 1992, Sedimentology and Stratigraphy of the Morrison Formation in Dinosaur National Monument, Utah and Colorado: Annual Report of the National Park Service (unpublished), contact #CA-1463-5-0001, in cooperation with the U.S. Geological Survey, 80 pp.

Chapter 6

1. Discovery of Giant Roaming Deep Sea Protist Provides New Perspective on Animal Evolution, The University of Texas at Austin press release, November 20, 2008, reporting on research published in M. V. Matz, T. M. Frank, N. J. Marshall, E. A. Widder, and S. Johnsen, 2008, Giant Deep-Sea Protist Produces Bilaterian-like Traces, *Current Biology*, 18 (23): 1849-1854.
2. D. Osorio, J. P. Bacon, and P. M. Whitington, 1997, The evolution of arthropod nervous systems, *American Scientist,* 85 (3): 244-253.
3. C. Darwin,1872, *The Origin of Species*, 6th ed., New York: Collier Books, 308.
4. M. Denton, 1985, *Evolution: A Theory in Crisis*, Bethesda, MD: Adler & Adler, 190, citing A. S. Romer, 1966, *Vertebrate Paleontology*, 3rd ed., Chicago: University of Chicago Press, 347-396.

Chapter 7

1. R. Barnes, 1980, Invertebrate Beginnings, *Paleobiology*, 6: 365-370.
2. J. W. Valentine and D. H. Erwin, 1987, Interpreting Great Developmental Experiments: The Fossil Record, in R. A. Raff and E. C. Raff, eds., *Development as an Evolutionary Process,* New York: Alan R. Liss, 84.

Chapter 8

1. D. Shu, 2003, A paleontological perspective of vertebrate origin, *Chinese Science Bulletin,* 48: 725-35.
2. "The fishes are of ancient ancestry, having descended from an unknown free-swimming protochordate [a tunicate or lancelet] ancestor." C. P. Hickman, Jr., L. S. Roberts, and A. Larson, 1997, *Integrated Principles of Zoology,* New York: McGraw-Hill, 499.
3. "The evolutionary radiation of the chordates from their primitive relatives must have occurred rapidly within the Early Cambrian or possibly earlier." D. Prothero, 2004, *Bringing Fossils to Life*, New York: McGraw-Hill, 350.
4. M. A. Fenton and C. L. Fenton, 1958, *The Fossil Book: A Record of Prehistoric Life*, New York: Doubleday, 342.
5. M. J. Benton, 2005, *Vertebrate Palaeontology*, Malden, MA: Blackwell Science, 39.
6. B. Stahl, 1985, *Vertebrate History: Problems in Evolution*, New York: Dover Publications, Inc., 126.
7. G. T. Todd, 1980, Evolution of the Lung and the Origin of Bony Fishes: A Causal Relationship, *American Zoologist*, 26 (4): 757.
8. G. V. Lauder, 1997, Interrelationships of Fishes, *American Zoologist,* 37 (3): 325.
9. Strahler, 1987, 408.
10. Fossil footprints give land vertebrates a much longer history, Uppsala University press release, January 7, 2010, reporting research published in G. Niedzwiedzki, P. Szrek, K. Narkiewicz, M. Narkiewicz, and P. E. Ahlberg, 2010, Tetrapod trackways from the early Middle Devonian period of Poland, *Nature*, 463 (7277): 43-48.
11. J. A. Clack, 2002, *Gaining Ground: The Origin and Evolution of Tetrapods*, Bloomington, IN: Indiana University Press, 127.

Chapter 9

1. Fossils Suggest Earlier Land-Water Transition of Tetrapod, Duke University press release, April 17, 2009, reporting on research published in V. Callier, J. A. Clack, and P. E. Ahlberg, 2009, Contrasting Developmental Trajectories in the Earliest Known Tetrapod Forelimbs, *Science*, 324 (5925): 364-367.
2. A. Meyer and R. Zardoya, 2003, Recent Advances in the (Molecular) Phylogeny of Vertebrates, *Annual Review of Ecology, Evolution, and Systematics*, 34: 321. See update by J. S. Anderson and H. Sues, eds., 2007, *Major Transitions in Vertebrate Evolution*, Bloomington, IN: Indiana University Press.
3. Stahl, 1985, 268 and 276.
4. Hickman, Roberts, and Larson, 1997, 549.
5. E. H. Colbert, M. Morales, and E. C. Minkoff, 2001, *Colbert's Evolution of the Vertebrates*, New York: Wiley-Liss, 159.
6. Colbert, Morales, and Minkoff, 2001, 128.
7. Benton, 2005, 244.
8. Colbert, Morales, and Minkoff, 2001, 154.
9. Stahl, 1985, 336.
10. See D. E. Quick and J. A. Ruben, Cardio-pulmonary anatomy in theropod dinosaurs: Implications from extant archosaurs, *Journal of Morphology*, 270 (10): 1232-1246.
11. S. A. Austin, J. R. Baumgardner, A. A. Snelling, L. Vardiman, and K. P. Wise, 1994, Catastrophic Plate Tectonics: A Global Flood Model of Earth History, in R. E. Walsh, ed., *Proceedings of the Third International Conference on Creationism*, Pittsburg, PA: Creation Science Fellowship, Inc.
12. W. A. Hoesch and S. A. Austin, 2004, Dinosaur National Monument: Jurassic Park or Jurassic Jumble?, *Acts & Facts*, 33 (4).
13. S. A. Austin and W. A. Hoesch, 2006, Do Volcanoes Come in Super-Size?, *Acts & Facts,* 35 (8).
14. C. M. Faux and K. Padian, 2007, The opisthotonic posture of vertebrate skeletons: postmortem contraction or death throes?, *Paleobiology*, 33 (2): 201-206.
15. For example, see New Blow for Dinosaur-Killing Asteroid Theory, National Science Foundation press release, April 27, 2009; and J. E. Fassett, 2009, New Geochronologic and Stratigraphic Evidence Confirms the Paleocene Age of the Dinosaur-Bearing Ojo Alamo Sandstone and Animas Formation in the San Juan Basin, New Mexico and Colorado, *Palaeontologia Electronica*, 12.1.3A.
16. M. H. Schweitzer, J. L. Wittmeyer, J. R. Horner, and J. K. Toporski, 2005, Soft-Tissue Vessels and Cellular Preservation in *Tyrannosaurus rex*, *Science*, 307 (5717): 1952-1955.
17. H. Fields, May 2006, Dinosaur Shocker, *Smithsonian*, 50-55.
18. R. F. Service, 2009, 'Protein' in 80-Million-Year-Old Fossil Bolsters Controversial *T. rex* Claim, *Science*, 324 (5927): 280.
19. J. Vinther, D. E. G. Briggs, R. O. Prum, and V. Saranathan, 2008, The colour of fossil feathers, *Biology Letters*, 4 (5): 522-525.
20. M. Wardrop, 2009, Scientists draw squid using its 15-million-year-old fossilized ink, *Telegraph*, posted on telegraph.co.uk August 19, 2009.

Chapter 10

1. Darwin, 1872, 172.
2. S. J. Gould and N. Eldredge, 1977, Punctuated equilibria; the tempo and mode of evolution reconsidered, *Paleobiology*, 3 (2): 147.
3. E. Mayr, 2001, *What Evolution Is*, New York: Basic Books, 65.
4. R. N. Melchor, S. de Valais, and J. F Genise, 2002, Bird-like fossil footprints from the Late Triassic, *Nature,* 417 (6892): 936-938.
5. S. Chatterjee, 1991, Cranial Anatomy and Relationships of a New Triassic Bird from Texas, *Philosophical Transactions: Biological Sciences*, 332 (1265): 277-342.
6. A. Feduccia, T. Lingham-Soliar, and J. R. Hinchliffe, 2005, Do feathered dinosaurs exist? Testing the hypothesis on neontological and paleontological evidence, *Journal of Morphology*, 266 (2): 125.
7. Faux and Padian, 2007.
8. J. Hecht, 2005, Large mammals once dined on dinosaurs, *New Scientist*, 2482: 6.
9. Anderson and Sues, 2007, 153.
10. Colbert, Morales, and Minkoff, 2001, 392.
11. C. Darwin, 1859, *On the Origin of Species*, London: John Murray, 184.
12. R. Harrison and M. M. Bryden, eds., 1988, *Whales, Dolphins, and Porpoises*, London: Merehurst Press, 14.
13. Benton, 2005, 333.
14. Colbert, Morales, and Minkoff, 2001, 375.
15. J. Palmer, Deer-like fossil is a missing link in whale evolution, *New Scientist*, posted on newscientist.com December 19, 2007.
16. W. Li, 1997, *Molecular Evolution*, Sunderland, MA: Sinauer Associates Inc.
17. Romer, 1966, 297-298.
18. Stahl, 1985, 486-487.
19. Stahl, 1985, 489.
20. Colbert, Morales, and Minkoff, 2001, 392.
21. Colbert, Morales, and Minkoff, 2001, 394.
22. P. D. Gingerich, N. A. Wells, D. E. Russell, and S. M. Ibrahim Shah, 1983, Origin of Whales in Epicontinental Remnant Seas: New Evidence from the Early Eocene of Pakistan, *Science*, 220 (4595): 403-406.
23. Colbert, Morales, and Minkoff, 2001, 394.
24. J. G. M. Thewissen, E. M. Williams, L. J. Roe, and S. T. Hussain, 2001, Skeletons of terrestrial cetaceans and the relationship of whales to artidactyls, *Nature*, 413: 227-281.
25. Pakicetidae. Posted on the Northeastern Ohio Universities Colleges of Medicine and Pharmacy website at www.neoucom.edu/Depts/Anat/Pakicetid.html.
26. NEOUCOM Scientist Discovers Missing Link: Dr. Hans Thewissen identifies whales' four-footed ancestor, Northern Ohio Universities Colleges of Medicine and Pharmacy press release, December 20, 2007.
27. Stahl, 1985, 489.
28. *In Search of Human Origins Part One*, NOVA Transcript, PBS airdate June 3, 1997, posted on pbs.org.

Chapter 11

1. W. A. Shear, 1999, Millipedes, *American Scientist*, 87: 234.
2. C. Zimmer, 1993, Insects Ascendant, *Discover*, 14: 30.
3. H. Akahane, T. Furuno, H. Miyajima, T. Yoshikawa, and S. Yamamoto, 2004, Rapid wood silicification in hot spring water: An explanation of silicification of wood during the Earth's history, *Sedimentary Geology*, 169: 219-228.
4. R. C. Moore, 1958, *Introduction to Historical Geology*, 2nd ed., New York: McGraw-Hill, 401.

Appendix

1. F. Guterl, Evolution: Birds Do It, *Newsweek*, May 6, 2002.
2. D. Prothero, 2008, Evolution: What missing link?, *New Scientist*, 197 (2645): 35.
3. M. J. Benton, 2015, *Vertebrate Palaeontology,* 4th ed., Hoboken, NJ: Blackwell Publishing, 39.
4. J. Lu, D. M. Unwin, X. Jin, Y. Liu and Q. Ji, 2009, Evidence for modular evolution in a long-tailed pterosaur with a pterodactyloid skull, *Proceedings of the Royal Society B*, 276 (1673).
5. J. Hanken and B. K. Hall, eds., 1993, *The Skull, Volume 2: Patterns of Structural and Systematic Diversity*, Chicago: University of Chicago Press, 179.
6. Unless otherwise noted, all A. Meyer and R. Zardoya, 2003, Recent Advances in the (Molecular) Phylogeny of Vertebrates, *Annual Review of Ecology, Evolution, and Systematics*, 34: 311–38 quotations from Kathleen Hunt are taken from her Transitional Vertebrate Fossils FAQ, posted on talkorigins.org.
7. S. M. Stanley, 1981, *The New Evolutionary Timetable: Fossils, Genes, and the Origin of Species*, New York: Basic Books, 71.

8. B. Stahl, 1974, *Vertebrate History: Problems in Evolution*, New York: Dover Publications, 38 (emphasis added).
9. C. P. Hickman, Jr., S. L. Keen, A. Larson, D. J. Eisenhour, H. l'Anson, and L. S. Roberts, 2011, *Integrated Principles of Zoology*, 15th ed., New York: McGraw-Hill, 512 (emphasis added).
10. Benton, 2015, 10-11.
11. Ibid, 11.
12. Early family ties: No sponge in the human family tree, Ludwig-Maximilians-Universität München press release, April 2, 2009.
13. S. A. Miller and J. P. Harley, 2013, *Zoology*, 9th ed., New York: McGraw-Hill, 331.
14. F. Delsuc, H. Brinkmann, D. Chourrout and H. Philippe, 2006, Tunicates and not cephalochordates are the closest living relatives of vertebrates, *Nature*, 439 (7079): 965-968.
15. Miller and Harley, 2013, 352.
16. Benton, 2015, 124.
17. Miller and Harley, 2013, 353 (emphasis added).
18. Benton, 2015, 51 (emphasis added).
19. P. Janvier, 2002, *Early Vertebrates,* Oxford: Clarendon Press, 284. See also M. Coates, 2009, Palaeontology: Beyond the Age of Fishes, *Nature,* 458 (7237): 413-414.
20. Hanken and Hall, 1993, 175.
21. M. V. H. Wilson, G. F. Hanke, and T. Märss, 2007, Paired fins of jawless vertebrates and their homologies across the agnathan-gnathostome transition, in *Major Transitions in Vertebrate Evolution*, J. S. Anderson and H. Sues, eds., Bloomington, IN: Indiana University Press, 122.
22. Benton, 2015, 59 (emphasis added).
23. Janvier, 2002, 135.
24. Ibid, 147.
25. Stahl, 1974, 192 (emphasis added).
26. J. A. Long, 2011, *The Rise of Fishes,* Baltimore: Johns Hopkins University Press, 95.
27. G. S. Williams, A Listing of Fossil Sharks and Rays of the World, Foreword, posted on afn.org/~afn02877 June 1, 2003.
28. C. J. Douady, M. Dosay, M. S. Shivji and M. J. Stanhope, 2003, Molecular phylogenetic evidence refuting the hypothesis of Batoidea (rays and skates) as derived sharks, *Molecular Phylogenetics and Evolution*, 26 (2): 215-221.
29. S. Adnet and H. Cappetta, 2008, New fossil triakid sharks from the Eocene of Prémontré, France and comments on fossil record of the family, *Acta Paleontologica Polonica*, 53 (3): 433-448.
30. J. Mallatt and C. J. Winchell, 2007, Ribosomal RNA genes and deuterostome phylogeny revisited: more cyclostomes, elasmobranchs, reptiles, and a brittle star, *Molecular Phylogenetics and Evolution*, 43: 1005.
31. Douady et al, 2003.
32. R. A. Martin, Evolution of a Super Predator, Biology of Sharks and Rays, ReefQuest Centre for Shark Research website. Posted on elasmo-research.org.
33. Benton, 2015, 65, Figure 3.15.
34. M. Allaby, ed., 2014, *The Concise Dictionary of Zoology,* Oxford, UK: Oxford University Press, 133.
35. B. Schaeffer and M. Williams, 1977, Relationships of Fossil and Living Elasmobranchs, *American Zoologist*, 17 (2): 293-302.
36. Benton, 2015, 175 (emphasis added).
37. Stahl, 1974, 185 (emphasis added).
38. Ibid.
39. Benton, 2015, 175.
40. Ibid.
41. E. H. Colbert, M. Morales and E. C. Minkoff, 2001, *Colbert's Evolution of the Vertebrates,* Wilmington, DE: Wiley-Liss, 56 (emphasis added).
42. B. Schaeffer and M. Williams, 1977, Relationships of Fossil and Living Elasmobranchs, *American Zoologist*, 17 (2): 293-302.
43. C. J. Underwood, S. F. Mitchell and K. J. Veltcamp, 1999, Shark and ray teeth from the Hauterivian (Lower Cretaceous) of north-east England, *Palaeontology*, 42 (2): 287-302.
44. Underwood, Mitchell, and Veltcamp, 1999, 298.
45. J. A. F. Garrick, 1955, Studies on New Zealand Elasmobranchii, Part V, *Transactions and Proceedings of the Royal Society of New Zealand 1868-1961*, 83: 563.
46. K. Hunt, Transitional Fossils FAQ, posted on holysmoke.org/tran-icr.htm.
47. Stahl, 1974, 186.
48. Colbert, Morales, Minkoff, 2001, 56.
49. Stahl, 1974, 184.
50. Colbert, Morales, Minkoff, 2001, 56.
51. Benton, 2015, 175.
52. Ibid.
53. R. A. Martin, The Origin of Modern Sharks, Biology of Sharks and Rays, ReefQuest Centre for Shark Research website. Posted on elasmo-research.org.
54. Benton, 2015, 179.
55. Stahl, 1974, 187.
56. Ibid.
57. Wilson, Hanke, Märss, 2007, 122.
58. G. Ortí and C. Li, 2009, Phylogeny and Classification, B. G. M. Jamieson, ed., *Reproductive Biology and Phylogeny of Fishes (Agnathans and Bony Fishes)*, Enfield, NH: Science Publishers.
59. Benton, 2015, Figure 2.10.
60. Miller and Harley, 2013, 313.
61. Colbert, Morales, Minkoff, 2001, 8.
62. Hickman et al, 2011, 520.
63. Ibid.
64. Stahl, 1974, 126.
65. Intentionally left blank.
66. Colbert, Morales, Minkoff, 2001, 53.
67. M. Zhu, W. Zhao, L. Jia, J. Lu, T. Qiao and Q. Qu, 2009, The oldest articulated osteichthyan reveals mosaic gnathostome characters, *Nature*, 458 (7237): 469-474.
68. M. Coates, 2009, Palaeontology: Beyond the Age of Fishes, *Nature* 458 (7237): 413-414.
69. Stahl, 1974, 151 (emphasis added).
70. Benton, 2015, Box 3.4.
71. J. S. Anderson and H. Sues, 2007, *Major Transitions in Vertebrate Evolution*, Bloomington, IN: Indiana University Press, 140.
72. Stahl, 1974, 153.
73. Colbert, Morales, Minkoff, 2001, 74 (emphasis added).
74. Allaby, 2014, 120.
75. Benton, 2015, 71.
76. Hickman et al, 2011, Figure 24.18.
77. Benton, 2015, 185.
78. Stahl, 1974, 152.
79. Allaby, 2014, 84.
80. Ibid.
81. Colbert, Morales, Minkoff, 2001, 66.
82. Anderson and Sues, 2007, 161.
83. Allaby, 2014, 84.
84. Long, 2011, 145.
85. Colbert, Morales, Minkoff, 2001, 66.
86. Ibid, 89.
87. Benton, 2015, 182.
88. Colbert, Morales, Minkoff, 2001, 63, Figure 5.2.
89. Benton, 2015, 189-90.
90. Allaby, 2014, 565-66.
91. Stahl, 1974, 157.
92. P. E. Olsen, A. R. McCune, and K. S. Thomson, 1982, Correlation of the early Mesozoic Newark Supergroup by vertebrates, principally fishes, *American Journal of Science*, 282: 13. See also page 15.
93. M. Ridley, 1996, *Evolution, Second Edition*, Wilmington, DE: Wiley-Blackwell.

94. M. W. Browne, Biologists Debate Man's Fishy Ancestors, *New York Times*, March 16, 1993.
95. Colbert, Morales, Minkoff, 2001, 82.
96. Benton, 2015, 69.
97. Ibid, 73. (Lobe-finned fish are within the Sarcopterygii.)
98. J. A. Clack, 2012, *Gaining Ground: The Origin and Evolution of Tetrapods*, Bloomington, IN: Indiana University Press, 77.
99. Stahl, 1974, 148-149.
100. J. S. Levinton, 2001, *Genetics, Paleontology and Macroevolution*, Cambridge, UK: Cambridge University Press, 74.
101. Clack, 2012, 128.
102. M. Zhu, X. Yu, and P. Janvier, 1999, A primitive fossil fish sheds light on the origin of bony fishes, *Nature,* 397 (6720): 607-610.
103. Zhu et al, 2009.
104. V. Callier, Ancient fossil may rewrite fish family tree. *Science News*, posted on sciencemag.org January 12, 2015 (emphasis added).
105. K. Vandepoele, W. De Vos, J. S. Taylor, A. Myer and Y. Van de Peer, 2004, Major events in the genome evolution of vertebrates: Paranome age and size differ considerably between ray-finned fishes and land vertebrates, *Proceeding of the National Academy of Sciences*, 101 (6): 1638-1643.
106. Stahl, 1974, 148.
107. Allaby, 2014, 539 (emphasis added).
108. Clack, 2012, 98 (emphasis added).
109. Janvier, 2002, 268.
110. Clack, 2012, 95.
111. Colbert, Morales, Minkoff, 2001, 82.
112. Clack, 2012, 92.
113. D. Swift, 2002, *Evolution: Under the Microscope*, Stirling, Scotland: Leighton Academic Press, 266.
114. Clack, 2012, 94. The quote is in relation to a paper by D. E. Rosen, P. L. Forey, B. G. Gardiner, and C. Patterson, 1981, Lungfishes, tetrapods, paleontology, and plesiomorphy, *Bulletin of the American Museum of Natural History*, 167 (4): 167.
115. Benton, 2015, 109.
116. Hickman et al, 2011, 552.
117. Colbert, Morales, Minkoff, 2001, 81.
118. Allaby, 2014, 228.
119. E. Jarvik in H. Schultze and R. Cloutier, eds., 1996, *Devonian Fishes and Plants of Miguasha, Quebec, Canada*, Munich: Verlag Dr. Friedrich Pfeil, 288.
120. J. Hecht, 2007, The fishy origin of our fingers and toes, *New Scientist*, 2627: 14.
121. Clack, 2012, 79 (emphasis added). See also Figure 2.4.
122. Benton, 2015, 86.
123. T. Nakamura, A. R. Gehrke, J. Lemberg, J. Szymaszek, and N. H. Shubin, 2016, Digits and fin rays share common developmental histories, *Nature,* 537 (7619): 225-28.
124. Benton, 2015, 86.
125. Clack, 2012, 46.
126. Ibid (emphasis added).
127. Benton, 2015, 87.
128. G. Todd, 1980, Evolution of the Lung and the Origin of Bony Fishes: A Causal Relationship, *American Zoologist,* 26 (4): 757.
129. E. B. Daeschler, N. H. Shubin, and F. A. Jenkins, 2006, A Devonian tetrapod-like fish and the evolution of the tetrapod body plan, *Nature*, 440 (7085): 757-763.
130. N. Shubin, 2008, *Your Inner Fish: A Journey into the 3.5-Billion-Year History of the Human Body*, New York: Pantheon Books.
131. 'Fishpod' reveals origins of head and neck structures of first land animals, The University of Chicago Medical Center press release, October 15, 2008. See also J. P. Downs, E. B. Daeschler, F. A. Jenkins, Jr., and N. H. Shubin, 2008, The cranial endoskeleton of *Tiktaalik roseae*, *Nature*, 445 (7215): 925-929.
132. J. N. Wilford, Fossil Called Missing Link From Sea to Land Animals, *The New York Times*, April 7, 2006.
133. C. A. Boisvert, E. Mark-Kurik and P. E. Ahlberg, 2008, The pectoral fin of *Panderichthys* and the origin of digits, *Nature*, 456 (7222): 636-638.
134. West Australian fossil find rewrites land mammal evolution. Monash University press release, October 19, 2006, reporting on research published in J. A. Long, G. C. Young, T. Holland, T. J. Senden and E. M. G. Fitzgerald, 2006, An exceptional Devonian fish from Australia sheds light on tetrapod origins, *Nature*, 444 (7116): 199-202.
135. Long et al, 2006.
136. G. Niedzwiedzki, P. Szrek, K. Narkiewicz, M. Narkiewicz, and P. E. Ahlberg, 2010, Tetrapod trackways from the early Middle Devonian period of Poland, *Nature*, 463 (7277): 43-48 (emphasis added).
137. P. Janvier and G. Clement, 2010, Palaeontology: Muddy tetrapod origins, *Nature*. 463 (7277): 40-41 (emphasis added).
138. Clack, 2012. 216.
139. Stahl, 1974, 202, Figure 6.2.
140. Benton, 2015, Figure 4.2.
141. Ibid, 109.
142. Ibid, 111.
143. Hickman et al, 2011, 552.
144. Anderson and Sues, 2007, 187.
145. Benton, 2015, 113.
146. J. S. Anderson, R. R. Reisz, D. Scott, N. B. Fröbisch and S. S. Sumida, 2008, A stem batrachian from the Early Permian of Texas and the origin of frogs and salamanders, *Nature*, 453 (7194): 515.
147. Allaby, 2014, 651.
148. Benton, 2015, 105.
149. Ibid, 113.
150. Anderson and Sues, 2007, 187.
151. Ibid, 182.
152. K. V. Kardong, 2012, *Vertebrates: Comparative Anatomy, Function, Evolution*, 6th ed., New York: McGraw Hill, 107.
153. Colbert, Morales, Minkoff, 2001, 119.
154. Allaby, 2014, 651.
155. Benton, 2015, 112.
156. Clack, 2012, 267.
157. Benton, 2015, Box 4.1.
158. Four-legged fish an evolutionary mistake, posted on physorg.com September 3, 2005, reporting research published in P. E. Ahlberg, J. A. Clack, and H. Blom, 2005, The axial skeleton of the Devonian tetrapod *Ichthyostega*, *Nature*, 437 (7055): 137-140.
159. Clack, 2012, 185.
160. Ibid, 161.
161. J. A. Clack, P. E. Ahlberg, S. M. Finney, P. Dominguez Alonso, J. Robinson and R. A. Ketcham, 2003, A uniquely specialized ear in a very early tetrapod, *Nature*, 425 (6953): 65-69.
162. Allaby, 2014, 333.
163. Kardong, 2012, 110 (under Stem-Amniotes).
164. O. R. P. Bininda-Emonds, M. Cardillo, K. E. Jones, R. D. E. MacPhee, R. M. D. Beck, R. Grenyer, S. A. Price, R. A. Vos, J. L. Gittleman, and A. Purvis, 2007, The delayed rise of present-day mammals, *Nature*, 446 (7135): 507-512.
165. Anderson and Sues, 2007, 341. See also Y. I. Wolf, I. B. Rogozin, N. V. Grishin and E. V. Koonin, 2003, Genome trees and the tree of life, *Trends in Genetics*, 18 (9): 472-479.
166. Hickman et al, 2011, 621 (emphasis added).
167. Benton, 2015, 328.
168. Anderson and Sues, 2007, 153.
169. J. Roach, 2002, Earliest Known Ancestor of Placental Mammals Discovered, *National Geographic*, posted on news.nationalgeographic.com April 24, 2002.
170. Discovery of New Species of Mammal Sparks Research into the Explosive Rise of Present-Day Placental Mammals, Carnegie Museum of Natural History press release, June 20, 2007.
171. R. L. Cifelli, and B. M. Davis, 2013, Paleontology: Jurassic fossils and mammalian antiquity, *Nature,* 500 (7461): 160-161.

172. Benton, 2015, 330, Box 10.3.
173. K. A. Kermack, F. Mussett and H. W. Rigney, 1973, A problem in morganucodontid taxonomy (Mammalia), *Zoological Journal of the Linnean Society*, 53 (2): 157.
174. T. S. Kemp, 2005, *The Origin and Evolution of Mammals*, New York: Oxford University Press, 162.
175. E. Lawrence, 2011, *Henderson's Dictionary of Biology*, 15th ed., New York: Benjamin Cummings, 157.
176. C. Sidor, 2001, Simplification as a trend in synapsid cranial evolution, *Evolution,* 55 (7): 1420. Cynodonts were the "probable ancestors of mammals," according to T. Palmer, 1999, *Controversy: Catastrophism and Evolution*, New York: Kluwer Academic Publishers, 338.
177. Benton, 2015, 323.
178. Kemp, 2005, 79.
179. Benton, 2015, 141.
180. Allaby, 2014, 623.
181. R. L. Carroll, 1997, *Patterns and Processes of Vertebrate Evolution*. Cambridge, UK: Cambridge University Press, 397.
182. Benton, 2015, Box 5.3.
183. Colbert, Morales, Minkoff, 2001, 271.
184. Anderson and Sues, 2007, 345 (emphasis added).
185. Benton, 2015, Box 10.1
186. Kemp, 2005, Figure 3.24 (d).
187. Stahl, 1974, 410, Figure 9.7.
188. Allaby, 2014, 187.
189. Colbert, Morales, Minkoff, 2001, 282.
190. Benton, 2015, 327.
191. Stahl, 1974, 408-411.
192. K. P. Dial, N. Shubin, and E. L. Brainerd, 2015, *Great Transformations in Vertebrate Evolution*, Chicago: The University of Chicago Press, 205.
193. Kemp, 2005, 173-74.
194. Colbert, Morales, Minkoff, 2001, 288 (emphasis added).
195. Meyer and Zardoya, 2003, 313.
196. Allaby, 2014, 226 (emphasis added).
197. Anderson and Sues, 2007, 355.
198. Colbert, Morales, Minkoff, 2001, 280-81 (emphasis added).
199. Kemp, 2005, 176-78 (emphasis added).
200. Z.-X. Luo, Successive Diversifications in Early Mammalian Evolution, in Anderson and Sues, 2007, 355.
201. Allaby, 2014, 228 (emphasis added).
202. Stahl, 1974, 416 (emphasis added).
203. Benton, 2015, 340.
204. Ibid, 341.
205. Anderson and Sues, 2007, 367.
206. Ibid, 339.
207. Ibid, 371.
208. Colbert, Morales, Minkoff, 2001, 318.
209. Benton, 2015, 341.
210. Allaby, 2014, 508 (emphasis added).
211. Stahl, 1974, 417 (emphasis added).
212. Colbert, Morales, Minkoff, 2001, 318.
213. Ibid (emphasis added).
214. Benton, 2015, 341-42.
215. Colbert, Morales, Minkoff, 2001, 333.
216. Benton, 2015, 376.
217. Allaby, 2014, 124.
218. Stahl, 1974, 507.
219. Colbert, Morales, Minkoff, 2001, 415.
220. Kemp, 2005, 262.
221. Benton, 2015, 367.
222. Colbert, Morales, Minkoff, 2001, 419 (emphasis added).
223. Allaby, 2014, 187.
224. Ibid, 602.
225. Stahl, 1974, 507.
226. Allaby, 2014, 52.
227. Colbert, Morales, Minkoff, 2001, 420.
228. Stahl, 1974, 514-515 (emphasis added).
229. Benton, 2015, 372.
230. Colbert, Morales, Minkoff, 2001, 422.
231. J. R. Boisserie, F. Lihoreau, and M. Brunet, 2005, The position of Hippopotamidae within Cetartiodactyla, *Proceedings of the National Academy of Sciences*, 102 (5): 1537-1541.
232. Benton, 2015, 372.
233. Stahl, 1974, 512 (emphasis added).
234. Colbert, Morales, Minkoff, 2001, 420.
235. Ibid, 420.
236. Ibid, 421.
237. Stahl, 1974, 515.
238. Benton, 2015, Figure 10.35(c).
239. Allaby, 2014, 615.
240. Colbert, Morales, Minkoff, 2001, 421.
241. Benton, 2015, 369.
242. Stahl, 1974, 514-515 (emphasis added).
243. Allaby, 2014, 96.
244. Stahl, 1974, 520.
245. Benton, 2015, 371.
246. Allaby, 2014, 305.
247. Colbert, Morales, Minkoff, 2001, 428.
248. Stahl, 1974, 522.
249. Colbert, Morales, Minkoff, 2001, 428.
250. L. Chuan-Kuei, R. W. Wilson, M. R. Dawson, and L. Krishtalka, 1987, The origin of rodents and lagomorphs, in *Current Mammalogy, volume 1*, New York: Plenum Press, 97-108.
251. P. G. Cox and L. Hautier, 2015, *Evolution of the Rodents*, Cambridge, UK: Cambridge University Press, 393.
252. Ibid, 4.
253. J. Meng, Y. Hu, and C. Li, 2003, The Osteology of *Rhombomylus* (Mammalia, Glires): Implications for Phylogeny and Evolution of Glires, *Bulletin of the American Museum of Natural History*, 275: 1-247.
254. Benton, 2015, 385.
255. Cox and Hautier, 2015, 240.
256. Colbert, Morales, Minkoff, 2001, 369.
257. J. Helt, 2006, Jurassic 'beaver' is new fossil record, *New Scientist,* 189: 16.
258. Allaby, 2014, 263.
259. Kemp, 2005, 275.
260. Colbert, Morales, Minkoff, 2001, 352, Figure 24-1.
261. M. L. Weston, Lagomorpha, *The Canadian Encyclopedia*, posted on thecanadianencyclopedia.com.
262. Allaby, 2014, 335 (emphasis added).
263. Benton, 2015, 388.
264. Ibid, 385.
265. Cox and Hautier, 2015, 277.
266. Good Luck Indeed: 53 Million-year-old Rabbit's Foot Bones Found, *ScienceDaily,* posted on sciencedaily.com March 24, 2008.
267. Chuan-Kuei et al, 1987. See also F. S. Szalay, M. J. Novacek and M. C. McKenna, 1993, *Mammal Phylogeny*, vols. 1 & 2, New York: Springer-Verlag.
268. Colbert, Morales, Minkoff, 2001, 311.
269. P. D. Gingerich, 1977, Patterns of evolution in the mammalian fossil record, in A. Hallam, ed., *Patterns of Evolution as Illustrated by the Fossil Record*, Amsterdam, NY: Elsevier Scientific Pub. Co., 469-500.
270. E. L. Lundelius, T. Downs, E. H. Lindsay, H. A. Semken, R. J. Zakrzewski, C. S. Churcher, C. R. Harington, G. E. Schultz, and S. D. Webb, 1987, The North American Quaternary sequence, in M. O. Woodburne, ed., *Cenozoic Mammals of North America: Geochronology and Biostratigraphy*, Berkeley: University of California Press; and J. Chaline, 1983, Modalites, Rythmes, Mecanismes de L'Evolution Biologique: Gradualisme phyletique ou equilibres ponctues? Editions du Centre National de la Recherche

Scientifique, Paris (collection of symposium papers, most in French with English abstracts provided, some in English).
271. Colbert, Morales, Minkoff, 2001, 362.
272. Hippo Ancestry Disputed, University of Calgary Science News, March 20, 2009, reporting research published in J. H. Geisler and J. M. Theodor, 2009, Hippopotamus and whale phylogeny, *Nature*, 458 (7236): E1-E4.
273. J. Boisserie, F. Lihoreau, and M. Brunet, 2005, The position of Hippopotamidae within Cetartiodactyla, *Proceedings of the National Academy of Sciences*, 102 (5): 1537-1541.
274. Benton, 2015, 372 (emphasis added).
275. Colbert, Morales, Minkoff, 2001, 392.
276. Anderson and Sues, 2007, 393.
277. M. Denton, 2016, *Evolution: Still a Theory in Crisis*, Seattle, WA: Discovery Institute Press, 242.
278. W. Gray and D. Bellamy, 1993, *Coral Reefs and Islands*, Newton Abbot, Devon, UK: David & Charles, 98.
279. Benton, 2015, 373.
280. Ibid, 373.
281. Colbert, Morales, Minkoff, 2001, 396.
282. Ibid, 398.
283. Allaby, 2014, 380.
284. R. Carroll, 1997, *Patterns and Processes of Vertebrate Evolution*. Cambridge, UK: Cambridge University Press, 329.
285. M. A. O'Leary and J. Gatesy, 2008, Impact of increased character sampling on the phylogeny of Cetartiodactyla (Mammalia), *Cladistics*, 24 (4): 397-442.
286. J. G. M. Thewissen, E. M. Williams, L. J. Roe and S. T. Hussain, 2001, Skeletons of terrestrial cetaceans and the relationship of whales to artiodactyls, *Nature*, 413 (6853): 277-281.
287. P. D. Gingerich, 1983, Evidence for Evolution from the Vertebrate Fossil Record, *Journal of Geological Education*, 31 (2): 140-144.
288. P. D. Gingerich, N. A. Wells, D. E. Russell, and S. M. Ibrahim Shah, 1983, Origin of Whales in Epicontinental Remnant Seas: New Evidence from the Early Eocene of Pakistan, *Science*, 220 (4595): 403-406.
289. Benton, 2015, 373 (emphasis added).
290. Ibid, 372.
291. Colbert, Morales, Minkoff, 2001, 395, Figure 26-1.
292. Ibid, 394.
293. Benton, 2015, 372 (emphasis added).
294. C. de Muizon, 2001, Walking with whales, *Nature*, 413 (6853): 259-260.
295. Colbert, Morales, Minkoff, 2001, 394.
296. Benton, 2015, 373.
297. P. D. Gingerich, B. H. Smith, and E. L. Simons, 1990, Hind Limbs of Eocene *Basilosaurus*: Evidence of Feet in Whales, *Science*, 249 (4965): 154-157.
298. J. G. M. Thewissen, 1994, Phylogenetic aspects of Cetacean origins: A morphological perspective, *Journal of Mammalian Evolution*, 2 (3): 173.
299. Colbert, Morales, Minkoff, 2001, 396.
300. Stahl, 1974, 492 (emphasis added).
301. Kemp, 2005, 262.
302. Allaby, 2014, 148.
303. The "Timescale and Horse Family Tree" is a part of Kathleen Hunt's Fossil Horses FAQ, a companion file to her Transitional Vertebrate Fossils FAQ posted on talkorigins.org.
304. Kardong, 2012, 726.
305. Colbert, Morales, Minkoff, 2001, 459.
306. L. Orlando, J. L. Metcalf, M. T. Alberdi, M. Telles-Antunes, D. Bonjean, M. Otte, F. Martin, V. Eisenmann, M. Mashkour, F. Morello, J. L. Prado, R. Salas-Gismondi, B. J. Shockey, P. J. Wrinn, S. K. Vasil'ev, N. D. Ovodov, M. I. Cherry, B. Hopwood, D. Male, J. J. Austin, C. Hänni, and A. Cooper, 2009, Revising the recent evolutionary history of equids using ancient DNA, *Proceedings of the National Academy of Sciences*, 106 (51): 21754-21759.
307. Ibid.
308. Ibid.
309. Colbert, Morales, Minkoff, 2001, 459.
310. J. Weiner, 1994, *The Beak of the Finch: A Story of Evolution in Our Time*, New York, Knopf.
311. Orlando et al, 2009.
312. Colbert, Morales, Minkoff, 2001, 266.
313. Allaby, 2014, 352.
314. Tapir, 2008, *New World Encyclopedia*, posted on newworldencyclopedia.org.
315. Stahl, 1974, 496.
316. Benton, 2015, 378, Figure 10.41.
317. Ibid (emphasis added).
318. Tapir, 2008 (emphasis added).
319. Colbert, Morales, Minkoff, 2001, 471.
320. B. S. Ferrero and J. I. Noriega, 2007, A new upper Pleistocene tapir from Argentina: remarks on the phylogenetics and diversification of neotropical Tapiridae, *Journal of Vertebrate Paleontology*, 27 (2): 504-511.
321. W. B. Scott, 1941, The Perrisodactyla, in W. B. Scott and G. L. Jepsen, The Mammalian Fauna of the White River Oligocene, Part 5, *Transactions of the American Philosophical Society*, 28: 747-946.
322. Allaby, 2014, 307.
323. Colbert, Morales, Minkoff, 2001, 472-473.
324. Benton, 2015, 378.
325. Stahl, 1974, 499, Figure 9.46.
326. Ibid, 497.
327. Colbert, Morales, Minkoff, 2001, 473.
328. Allaby, 2014, 262.
329. Stahl, 1974, 523.
330. Benton, 2015, Figure 10.37
331. Allaby, 2014, 262.
332. Colbert, Morales, Minkoff, 2001, 433.

Selected Bibliography

Akahane, H. T., Furuno, H. Miyajima, T. Yoshikawa, and S. Yamamoto, 2004, Rapid wood silicification in hot spring water: An explanation of silicification of wood during the Earth's history, *Sedimentary Geology*, 169: 219-228.

Allaby, M., 1992, *The Concise Oxford Dictionary of Zoology*, Oxford: Oxford University Press.

Anderson, J. S., and H. Sues, eds., 2007, *Major Transitions in Vertebrate Evolution*, Bloomington, IN: Indiana University Press.

Arct, M. J., 1991, Dendroecology in the Fossil Forests of the Specimen Creek Area, Yellowstone National Park, Ph.D. dissertation, Loma Linda University, Loma Linda, CA.

Austin, S. A., and W. A. Hoesch, 2006, Do Volcanoes Come in Super-Size?, *Acts & Facts*, 35 (8).

Austin, S. A., J. R. Baumgardner, A. A. Snelling, L. Vardiman, and K. P. Wise, 1994, Catastrophic Plate Tectonics: A Global Flood Model of Earth History, in R. E. Walsh, ed., *Proceedings of the Third International Conference on Creationism*, Pittsburg, PA: Creation Science Fellowship, Inc.

Barnes, R., 1980, Invertebrate Beginnings, *Paleobiology*, 6: 365-370.

Benton, M. J., 2005, *Vertebrate Palaeontology*, Malden, MA: Blackwell Science.

Chatterjee, S., 1991, Cranial Anatomy and Relationships of a New Triassic Bird from Texas, *Philosophical Transactions: Biological Sciences*, 332 (1265): 277-342.

Clack, J. A., 2002, *Gaining Ground: The Origin and Evolution of Tetrapods*, Bloomington, IN: Indiana University Press.

Colbert, E. H., M. Morales, and E. C. Minkoff, 2001, *Colbert's Evolution of the Vertebrates*, 5th ed., New York: Wiley-Liss.

Darwin, C., 1872, *The Origin of Species*, 6th ed., New York, NY: Collier Books.

Denton, M., 1985, *Evolution: A Theory in Crisis*, Bethesda, MD: Adler & Adler.

Donoghue, P. C. J., 2007, Palaeontology: Embryonic identity crisis, *Nature*, 445 (7124): 155-156.

Englmann, G., and G. Callison, 1998, Mammalian faunas of the Morrison Formation, *Modern Geology*, 23: 343-380.

Faux, C. M., and K. Padian, 2007, The opisthotonic posture of vertebrate skeletons: postmortem contraction or death throes?, *Paleobiology*, 33 (2): 201-206.

Feduccia, A., T. Lingham-Soliar, and J. R. Hinchliffe, 2005, Do feathered dinosaurs exist? Testing the hypothesis on neontological and paleontological evidence, *Journal of Morphology*, 266 (2): 125-166.

Fenton, M. A., and C. L. Fenton, 1958, *The Fossil Book: A Record of Prehistoric Life*, New York: Doubleday.

Fields, H., May 2006, Dinosaur Shocker, *Smithsonian*, 50-55.

Futuyma, D., 1983, *Science on Trial*, NewYork: Pantheon Books.

Gingerich, P. D., N. A. Wells, D. E. Russell, and S. M. Ibrahim Shah, 1983, Origin of Whales in Epicontinental Remnant Seas: New Evidence from the Early Eocene of Pakistan, *Science*, 220 (4595): 403-406.

Gould, S. J., 1977, Evolution's Erratic Pace, *Natural History*, 86 (6).

Gould, S. J., 1991, Not Necessarily a Wing, *Bully for Brontosaurus*, New York: W. W. Norton & Co., 139-151 (originally published October 1985 as Not Necessarily a Wing, *Natural History*, 94: 12-25).

Harrison, R., and M. M. Bryden, eds., 1988, *Whales, Dolphins and Porpoises*, London: Merehurst Press.

Hecht, J., 2005, Large mammals once dined on dinosaurs, *New Scientist*, 2482: 6.

Hickman, Jr., C. P., L. S. Roberts, and A. Larson, 1997, *Integrated Principles of Zoology*, New York: McGraw-Hill.

Hoesch, W. A., and S. A. Austin, 2004, Dinosaur National Monument: Jurassic Park or Jurassic Jumble?, *Acts & Facts*, 33 (4).

In Search of Human Origins Part One, NOVA Transcript, PBS airdate June 3, 1997, posted on pbs.org.

Koonin, E. V., 2007, The Biological Big Bang model for the major transitions in evolution, *Biology Direct*, 2: 21.

Lauder, G. V., 1997, Interrelationships of Fishes, *American Zoologist*, 37 (3): 325.

Lewontin, R., 1997, Billions and Billions of Demons: Review of *The Demon-Haunted World* by Carl Sagan, *The New York Review of Books*, 44 (1): 31.

Li, W., 1997, *Molecular Evolution*, Sunderland, MA: Sinauer Associates Inc.

Mayell, H., "Mummified" Dinosaur Discovered In Montana, *National Geographic News*, posted on news.nationalgeographic.com October 11, 2002.

Mayr, E., 2001, *What Evolution Is*, New York: Basic Books.

Melchor, R. N., S. de Valais, and J. F Genise, 2002, Bird-like fossil footprints from the Late Triassic, *Nature*, 417 (6892): 936-938.

Meyer, A., and R. Zardoya, 2003, Recent Advances in the (Molecular) Phylogeny of Vertebrates, *Annual Review of Ecology, Evolution, and Systematics*, 34: 321.

Moore, R. C., 1958, *Introduction to Historical Geology*, 2nd ed., New York: McGraw-Hill.

Morris, J. D. 2007, *The Young Earth*, Dallas, TX: Institute for Creation Research

Noble, E. R., G. A Noble, G. A. Schad, and A. J. MacInnes, 1989, *Parasitology: The Biology of Animal Parasites*, 6th ed., Philadelphia, PA: Lea & Febiger.

Numbers, R. L., 1992, *The Creationists*, New York, NY: Alfred A. Knopf.

Osorio, D., J. P. Bacon, and P. M. Whitington, 1997, The evolution of arthropod nervous systems, *American Scientist,* 85 (3): 244-253.

Palmer, J., Deer-like fossil is a missing link in whale evolution, *New Scientist*, posted on newscientist.com December 19, 2007.

Prothero, D., 2004, *Bringing Fossils to Life*, New York: McGraw-Hill.

Quick, D. E., and J. A. Ruben, Cardio-pulmonary anatomy in theropod dinosaurs: Implications from extant archosaurs, *Journal of Morphology*, 270 (10):1232-1246.

Ridley, M., 1981, Who doubts evolution? *New Scientist,* 90: 832.

Romer, A. S., 1966, *Vertebrate Paleontology*, 3rd ed., Chicago: University of Chicago Press.

Ruse, M., Saving Darwin from the Darwinians, *National Post*, May 13, 2000, B-3.

Schweitzer, M. H., J. L. Wittmeyer, J. R. Horner, and J. K. Toporski, 2005, Soft-Tissue Vessels and Cellular Preservation in *Tyrannosaurus rex, Science*, 307 (5717): 1952-1955.

Service, R. F., 2009, 'Protein' in 80-Million-Year-Old Fossil Bolsters Controversial *T. rex* Claim, *Science*, 324 (5927): 280.

Shear, W. A., 1999, Millipedes, *American Scientist*, 87: 234.

Shu, D., 2003, A paleontological perspective of vertebrate origin, *Chinese Science Bulletin,* 48: 725-35.

Stahl, B., 1985, *Vertebrate History: Problems in Evolution*, New York: Dover Publications, Inc.

Stanley, S., 1977, *Macroevolution*, San Francisco, CA: Freeman.

Strahler, A. N., 1987, *Science and Earth History: The Evolution/Creation Controversy*, Buffalo: Prometheus Books.

Thewissen, J. G. M., E. M. Williams, L. J. Roe, and S. T. Hussain, 2001, Skeletons of terrestrial cetaceans and the relationship of whales to artidactyls, *Nature*, 413: 227-281.

Todd, G. T., 1980, Evolution of the Lung and the Origin of Bony Fishes: A Causal Relationship, *American Zoologist*, 26 (4): 757.

Turner, C., and F. Peterson, 1992, Sedimentology and Stratigraphy of the Morrison Formation in Dinosaur National Monument, Utah and Colorado: Annual Report of the National Park Service (unpublished), contact #CA-1463-5-0001, in cooperation with the U.S. Geological Survey.

Valentine, J. W., and D. H. Erwin, 1987, Interpreting Great Developmental Experiments: The Fossil Record, in R. A. Raff and E. C. Raff, eds., *Development as an Evolutionary Process*, New York: Alan R. Liss.

Vardiman, L., A. A. Snelling, and E. F. Chaffin, eds., 2000, *Radioisotopes and the Age of the Earth*, vol. 1, El Cajon, CA: Institute for Creation Research, and St. Joseph, MO: Creation Research Society.

Vinther, J., D. E. G. Briggs, R. O. Prum, and V. Saranathan, 2008, The colour of fossil feathers, *Biology Letters*, 4 (5): 522-525.

Wardrop, M., 2009, Scientists draw squid using its 15-million-year-old fossilized ink, *Telegraph*, posted on telegraph.co.uk August 19, 2009.

Wilson, E. O., 1982, Toward a Humanistic Biology, *The Humanist*, 42: 40.

Zimmer, C., 1993, Insects Ascendant, *Discover*, 14: 30.

Photo and Illustration Credits:

A=All, T=Top, M=Middle, B=Bottom, L=Left, R=Right

Canopy Ministries: 113R

Creation and Earth History Museum: 14

Creationism.org: 72B

Harold Coffin: 118R

field.ca: 51

Fotolia: 47A (except MR), 49R, 57, 58B, 71A, 105, 114TL, 114BL, 121TR, 121MR, 121BR

Jens L. Franzen, Philip D. Gingerich, Jörg Habersetzer1, Jørn H. Hurum, Wighart von Koenigswald, B. Holly Smith (Wikipedia): 110

Getty: 65

Heritage Auctions, Inc.: 8, 12, 16, 20, 24, 50, 54, 55, 56, 59B, 60T, 61TL, 61R, 67, 68, 72T, 74L, 76B, 80, 102, 112, 113BL, 114BR, 115A, 120R, 122

iStock: 34M, 59TL, 60BL, 106R, 114TR, 120R

Ian Juby: 117A

Dr. John Morris: 33R, 36A, 37A, 52T, 66, 78, 82BR, 89, 90A, 91, 94, 116A, 118L, 126

NASA: 38 (satellite image), 39 (satellite image), 42, 47MR

North Dakota Geological Survey: 49L

NPS: 26, 30, 60BR, 74R, 82BL, 93, 120L, 121TL, 121ML, 121BL

Public Domain: 73, 82T, 106L, 108, 109

Raymond Spekking: 61BL, 84, 86

Joe Taylor: 31

University of Kansas: 59TR

Verisimilus at en.wikipedia: 32, 45

Susan Windsor: 10A, 28, 34A (except M), 41, 44, 53, 57B, 58T, 62, 64, 69, 76T, 77, 88, 97, 98, 99A, 101A, 103

Index